AF864199

THE NATURE OF THE PAGE

MATERIAL TEXTS

Series Editors

Roger Chartier
Joseph Farrell
Anthony Grafton
Leah Price
Peter Stallybrass
Michael F. Suarez, S.J.

A complete list of books in the series
is available from the publisher.

The NATURE *of the* PAGE

Poetry, Papermaking,
and the Ecology of Texts
in Renaissance England

Joshua Calhoun

PENN

University of Pennsylvania Press
Philadelphia

Copyright © 2020 University of Pennsylvania Press

All rights reserved. Except for brief quotations used for purposes of review
or scholarly citation, none of this book may be reproduced in any form
by any means without written permission from the publisher.

Published by
University of Pennsylvania Press
Philadelphia, Pennsylvania 19104-4112
www.upenn.edu/pennpress

Printed in the United States of America on acid-free paper

1 3 5 7 9 10 8 6 4 2

A Cataloging-in-Publication record is available from the Library of Congress
ISBN 978-0-8122-5189-0

For
Misty Arden, my Chickadee
"Haply I think on thee . . ."

CONTENTS

PREFACE

Beginnings

> Tongues in trees, books in the running brooks,
> Sermons in stones.
>
> —Duke Senior in William Shakespeare,
> *As You Like It*, 2.1.16–17

This book tells a story about paper in Renaissance England—about what it was elementally, and about what it was not; about what a page of paper did, what it was made to do, and what it would not do; about what it made representable and unrepresentable, recordable and revisable, preservable and destructible. It is a story about recording so much of what we call history on sloshed-together plant fibers. For most of the history of printing, paper was made primarily from recycled rags, so this is also a story about using tattered ship sails and worn-out clothes to tell new stories about the past, about the plant fibers used to make those textiles that were eventually used to make texts, and about the plant fibers that frustrated papermakers' best attempts to replace scarce natural resources with abundant natural resources. Paper, in the story this book tells, is a marvelous but flawed protagonist, the product of nature and culture, of nonhuman and human agency. This story about human ideas recorded on plants is also an environmental story about the ecology of paper and about the ecosystems in which poets and plants can become (and un-become) Renaissance literature. And because plants, like humans, are defenseless against "Time's scythe," this is also a story about corruption—corruption and replication and the desperate hope that we can out-replicate the thing we love so as to preserve it from decay.[1]

We have, by and large, taken for granted the ecologies that allow, disallow, and alter the storage and transmission of ideas. We overlook not only the nature of handmade pages, but also the nature of the electronic screens on which we access digital reproductions of those pages and record our own ideas. Portions of this book, especially ideas that came at moments when keyboard and screen or pen and paper were not manageable, were first recorded on an iPhone, a now ubiquitous communication device that, in its earliest versions, was made with "toxins" such as arsenic, beryllium, lead, and mercury.[2] In 2015, Apple Inc.'s new take-back initiatives aimed at recycling "finite resources" recovered nearly 200,000 pounds of cobalt, more than 2,000 pounds of gold, and more than 4.5 million pounds of aluminum from old iPhones.[3] Though we may not have such precise statistics for natural resource usage in sixteenth- and seventeenth-century bookmaking, we know that, like smartphones, Renaissance books were made from and with finite resources. They were also made with visible, recognizable traces of ecological matter: recycled clothes, slaughtered animals, felled trees.

The Nature of the Page draws attention to the plant, animal, and mineral materials employed by human creatures, who seem to have a unique need to externalize cognition and memory, creatures whose minds are bursting with ideas that they want to transfer to some savable, shareable format. This study traces the plant fibers found in handmade papers through the late 1800s, when recycled rags were replaced by living trees as the stuff that stories, like this very book, are made on. My focus is especially on the ways in which the production and use of handmade paper have influenced and been influenced by global resource availability in an age of burgeoning exploration and colonization and natural resource extraction. Eating, we know, has human advantages and ecological consequences. Agriculture has profoundly altered our planet. Writing and reading, too, have human advantages and ecological consequences, and on a scale that we have not yet honestly acknowledged in our stories about book history or fully recognized in our studies of environmental history.

Acknowledging and engaging with the ecology of media in other periods and places, this book focuses on a particular medium, paper, in a particular time and place, Renaissance England. Yet the questions I ask of early handmade paper might also be just as productively asked of millennia-old Eastern palm-leaf books or medieval scrolls or Victorian headstones or junk mail or the newest iPhone. They might be distilled into three questions that guide this work: (1) How has scarcity of nonhuman matter altered human communication? (2) How have humans creatively imagined or reimagined the textual possibilities available to them in a given ecosystem? (3) How has human communication been altered by the corruptibility of the nonhuman matter used to make texts? Scarcity. Possibility. Corruptibility. These three ecopoetic negotiations—as pertinent to twenty-first-century ebooks as to sixteenth-century books on handmade paper—guide *The Nature of the Page*'s narrative.

Paper mills required rivers, and I think it is appropriate that two rivers have shaped my own understanding of the nature of the page. I grew up near the source of the Hudson River in the patchwork wilderness of the Adirondack Park, a six-million-acre area—larger than Yellowstone, Everglades, Glacier, and Grand Canyon National Parks combined—of which roughly half is private land (villages, businesses, farms, etc.) and the other half is forest preserve that has been designated "forever wild."[4] With age and reading and train trips down the Hudson River to access rare books in archival libraries came the realization that the Adirondacks were not a sovereign island of wilderness but were, and in many ways continue to be, New York City's hinterland. My journeys through the watersheds and river valleys between the Adirondacks and archival libraries have shaped this work and have left me unable to think about technology and progress without also asking about wilderness and landscape.

Now in Madison, Wisconsin, I write these words less than fifty miles from where Aldo Leopold once stood on the edge of the Wisconsin River looking at a piece of driftwood and jotting down observations that would, with enough paper and time, become these lines in *A Sand County Almanac*:

> The spring flood brings us more than high adventure; it brings likewise an unpredictable miscellany of floatable objects pilfered from upriver farms. . . . Each old board has its own individual history, always unknown, but always to some degree guessable from the kind of wood, its dimensions, its nails, screws, or paint, its finish or the lack of it, its wear or decay. One can even guess, from the abrasion of its edges and ends on sandbars, how many floods have carried it in years past.[5]

Here Leopold offers what we might now recognize as a material culture reading of lumber that is invested not only in political and cultural systems, but also, and especially, in ecosystems. Drawing attention to "biotic interactions between people and land," Leopold claims that the riparian lumber is "not only a collection of personalities, but an anthology of human strivings in upriver farms and forests."[6] The person who understands human strivings in upriver farms and forests has a kind of ecological literacy to reconstruct the history of a piece of driftwood, Leopold claims. The driftwood serves, in his account, as "a kind of literature" that might be "taught on campuses," a record of the past that is accessible and available to be "read at will."[7] The language he uses to describe the ecological readings is that of eager curiosity tempered by sensible humility: the history of a board, Leopold claims, is "always unknown, but always to some degree guessable from the kind of wood, its dimensions, its nails, screws, or paint, its finish or lack of it, its wear or decay."[8] This language of discovery, of the unknown but to some degree guessable, of drawing on imperfect expertise in an attempt to make the ecologies of media more legible, aptly describes the project that *The Nature of the Page* undertakes.

I quote these lines from a specially issued Leopold Pines Edition of *A Sand County Almanac*, an edition that, in 2007, was "printed on paper made from pines planted by Aldo Leopold and his family in the 1930s and '40s."[9] In a foreword to the edition, Nina Leopold Bradley, A. Carl Leopold, and Estella B. Leopold, who, as children, helped to plant the very pines from which the volume's paper was made, write of the "happy continuity" of their father's "precious pine trees becoming the medium, the paper, on which we print his moving words."[10] Reading Leopold reading driftwood on pages made from the pine trees he planted with his children is, for me, quite a lot like holding Renaissance books printed or written on handmade paper, objects whose human ecology repeatedly interrupts the text on the page and insists on being read. History is mediated by nature; nature is mediated by history. The Leopold Pines Edition is one of many examples I cite in this book where the nature of the medium, paper, intertwines with the message it carries in complex ways that are mostly unknown, but that are at least partially guessable. *The Nature of the Page*, then, tells a story about textual habits grounded in and supplied by ecological habitats. It is a story about the plants, animals, and minerals on which the poetry of Shakespeare and his contemporaries was stored and then transmitted across time and space. And it is a story, in the end, about the ecological resources we use now in our attempts to preserve the poetry we have received from the past on thin, pliable, handcrafted leaves of organic matter.

INTRODUCTION

Toward an Ecology of Texts

> What must matter to the environmental humanities is how texts are entangled with and address the larger processes by which societies conceptualize and manage their environment.
>
> —Hannes Bergthaller et al., "Mapping Common Ground: Ecocriticism, Environmental History, and the Environmental Humanities"

In "The Book," a short poem published in 1655, Henry Vaughan observes that his Bible's paper is made from flax plants and that its binding is made from the wood of a tree covered with the skin of a "harmless beast." What happens to a Bible, Vaughan wonders, on the day of resurrection, when all things are restored to their perfect forms? Does the Bible revert to its vegetal, bestial origins? Vaughan's poem raises questions about the earthly matter from which media are made, about how nonhuman objects in bookish format persist, interrupt, and alter the words they are made to record. Embedded in the pages of Renaissance texts are legible ecologies that record the environmental negotiations of people and things, of humans, humanists, and nonhumans. I return to "The Book" in Chapter 2, but here I introduce the poem in summary to call attention to its intriguing mode of reading. Vaughan's reading of a book, like the reading of driftwood discussed in the Preface, draws on what is known in order to imagine the material history of the text in his hands. Like Leopold, Vaughan acknowledges his own role in the narrative as the one who is holding the object and envisioning its "unknown, but to some degree guessable" past. But then Vaughan's reading takes a sudden, surprising turn: he pivots around the object to look from its past creation to its future destruction. Both book and reader intersect in a moment of flux, and Vaughan knows that the nonhuman things he holds, gathered in the past into the form of a book, must be scattered again in the future.

The Nature of the Page takes up and extends the Janus-faced mode of reading past and future that is modeled in "The Book." The readings I offer here attempt to account for what the text is, at the moment and place of human intersection, but also to draw out the past and future lives of the nonhuman things we recognize, metonymically, as texts. In an influential essay that laid the groundwork for future scholarship on material texts, Margreta de Grazia and Peter Stallybrass call attention to the fact that a Renaissance text is "a provisional state in the circulation of matter."[1] Literary critics and book historians have, in recent years, drawn attention to the provisionality of texts as they circulate in society, but *The Nature of the Page* extends the conversation about provisions and provisionality to its literal roots. Drawing attention to ecological provisionality in its many senses—temporary, preliminary, provident, conditional—I adjust the depth of field so that nonhuman histories (and futures) of material texts come into view and become legible. The words on the pages of this book, printed with soy-based ink on recycled tree pulp, engage with and draw attention to the textual forms of ship sails, hemp, flax, ink blots, animal glue, human hair, and fungi.[2] It is easy to imagine humans as the point of origination, as if the materials used to make paper, to make ink, to make printing type, and so on simply existed in abundance, waiting to be harvested. In reality, ecological availability and scarcity make certain kinds of human records possible; at the moment those possibilities are realized and integrated into textual forms, textual corruption and disintegration begin. The mode of reading modeled in "The Book" and pursued on a larger scale here inspires a more ecologically deterministic, but also a more poetic past and future for those endlessly intriguing sites of humanistic and environmental negotiation we call "texts." At the heart of this work, or more appropriately, at the headwaters, is a fascination with the ecopoetic motif of textual negotiation, with moments when scarcity, possibility, or corruptibility interrupts writers and readers.

One finds textual negotiation as an ecopoetic motif woven throughout English Renaissance literature: over and over again, sixteenth- and seventeenth-century writers express frustration, surprise, impatience, and inventiveness as they confront the affordances of various ecological materials in textual form. At times the animal, plant, and/or mineral materials used to make texts are taken to be affective, as when Thomas Dekker and George Wilkins describe a writer so filled with vitriol that his words should appear on "paper made of the filthy linnen rags that had beene wrapt about the infected and vlcerous bodyes of beggers, that had dyed in a ditch of the pestilence."[3] At times the materials are the subject of poetic conceits, as when Vaughan, in the example above, recognizes the former states of paper (seed, flax plant, and clothing) not only as a fact of textual production, but also as an inspiration for poetry about a material and metaphysical conundrum. At times the materials are hidden behind the very metaphors and poetic tropes that they have inspired, as when so many English poets, drawing on Petrarch, who was drawing on Virgil, refer to their verses as "scattered leaves." Quite often, earthy materials are blamed for their inability to do justice to a subject, as when the speaker of Shakespeare's *Sonnets* asks, "What's in the brain that ink may character . . . ?"[4] At other times, the materials are blamed for their seeming resistance: when Philip Sidney as Astrophil struggles to find "fit words to paint the blackest face of woe," seeking inspiration in "others' leaves" and biting his "truant" quill pen, he struggles with words but partly blames a feather.[5] By the poem's end, Sidney claims his muse has inspired him to the point that he feels capable of mastery over his materials. He will look in his heart and write. It is a nice fiction, but as Adam Smyth observes when discussing the theme of ineffability in William Strode's work, "Books are not disembodied conveyers of meaning but papers, ink marks, sheets, and strings."[6] And we could add, adjusting the depth of field, flax seeds and stalks, and rain, and sunlight, and the decayed bodies of things as nutrient-rich soil.

Recognizing that scrolls, and books, and tablet screens are sites of ecological negotiation, that the acts of writing and reading are also meaningful and entangled modes of dwelling, *The Nature of the Page* draws literary criticism into conversation with two generative fields of scholarly study that are not typically linked: book history and the environmental humanities. Both fields are expansive and difficult to define precisely. For that matter, both fields continue to debate their own "fieldness." Robert Darnton, addressing concerns about "interdisciplinarity run riot" as early as the 1980s suggested that book history "looks less like a field than a tropical rain forest"; Darnton's biogeographic metaphor now fits the environmental humanities as well or better than it does book history.[7] Hannes Bergthaller and colleagues refer to the environmental humanities as a "metadiscipline or superfield," and Ursula K. Heise labels it an "interdisciplinary matrix."[8] In their introduction to the first issue of the journal *Environmental Humanities*, Deborah Bird Rose and colleagues suggest that the environmental humanities gained momentum as "what have traditionally been termed 'environmental issues' have been shown to be inescapably entangled with human ways of being in the world."[9] This book makes these inescapable entanglements more visible and more legible by calling attention to "the world of things themselves" in which humans lived, and moved, and made their being known through different kinds of animal, vegetable, and mineral memorials.[10] A central claim of this work is that the story of paper is as much an environmental story as it is a bibliographical story. The story of paper is an ecopoetic story, too, for making poetry in the world means not only finding the right words, but finding the right matter to convey those words. Focusing on early handmade paper as a case study, *The Nature of the Page* draws out three strands, named in the book's subtitle, that tell us much about the environmental histories of textual negotiation: poetry, papermaking, and the ecology of texts.

If, as with book history, scholars do not always agree when, precisely, the environmental humanities began and which disciplines and disciplinary approaches have most contributed to its growth, few would argue with the claim that "two fields that have embraced the environmental humanities with particular fervour [are] ecocriticism and environmental history."[11] *The Nature of the Page* is a project with roots in both approaches, as will be evident throughout the book. As an epigraph to this Introduction, I cite "Mapping Common Ground: Ecocriticism, Environmental History, and the Environmental Humanities," a pivotal essay collaboratively authored by a broadly international group of ten literature, history, and geography scholars. In the essay, the authors make an impassioned argument for the value of scholarly heterodoxy, reflexivity, and experimentation, all grounded in an understanding that literary analysis can best be "conducted in a mode attuned to social practices of environing (rather than taking the existence of 'the environment' as a given)."[12] Grounding media and media-making in a landscape (or multiple landscapes as texts travel through time and space) allows different, more nuanced, and more entangled considerations to filter into ongoing conversations about book history and the environmental humanities. Such an approach calls attention to the influence of habitat on textual habits and allows us to understand the place-based constraints of book technologies.

Paper and Place

In Book 2 of Thomas More's *Utopia*, the fictional narrator Raphael Hythloday outlines the history of the book on the island of Utopia, an island whose name emphasizes its placelessness.[13] Hythloday claims he arrived on the island of nowhere with "a great many Books" including works of Plato and Aristotle, Plutarch, Lucian, Aristophanes, Euripides, Sophocles, Thucydides, and Herodotus. One of his companions had copies of Hippocrates and Galen. Hythloday also had an incomplete copy of Theophrastus's work on plants (incomplete thanks to a page-tearing monkey on board the ship). These books intrigued the Utopians because they were printed (not handwritten) on paper, a substance made from macerated plant fibers. According to Hythloday, "While we were showing them the Aldine editions of various books, we talked about papermaking and type-cutting, though without going into details, for none of us had

had any practical experience. But with great sharpness of mind they immediately grasped the basic principles. While previously they had written only on vellum, bark and papyrus, they now undertook to make paper and print with type."[14] An early nineteenth-century translation sums it up this way: "Formerly they wrote only on parchment, reeds, or the bark of trees. Now they have established paper-manufactures and printing-presses."[15] After some trial and error with these new technologies, Hythloday says, they "mastered both arts . . . reprinting each [of the Aldine editions] in thousands of copies."[16] In Hythloday's narrative, then, resourceful humans on one side of the world see the handiwork of resourceful humans from the other side of the world and, with a bit of practice, they reproduce it. No mention is made of the plants or plant-based fibers that must have supplied so much papermaking, nor would it make much difference in a fictional world. If Hythloday had said that the Utopians ultimately learned to make paper out of Japanese mulberry, it would be useless to question the claim. Mulberry trees may be written into existence on a fictional island. A Utopian ecosystem that is nowhere cannot lack natural resources.

Hythloday's placeless history of the book in the fictional environment of Utopia is much like the book history narratives we have tended to tell, as is best demonstrated by Darnton's influential schematic diagram "The Communications Circuit" (see Figure 1). In the diagram, only outputs—human labor and the texts generated by that labor—are mapped; the nonhuman inputs are assumed. Like a book in Utopia, books in Darnton's circuit exist as if in a human vacuum rather than in an ecosystem of plants, animals, and minerals whose availability or scarcity dictates what a text can be at any given time and in any given place.[17] Leslie Howsam praises Darnton's model for its "emphasis on human agency in the making and use of books," calling attention to a sociological shift in book history, one that brought human labor into focus.[18] This important shift in book history was best and most famously sounded by D. F. McKenzie in *Bibliography and the Sociology of Texts*. Redefining bibliography as "the study of the sociology of texts," McKenzie insisted that humans must be a part of the stories that scholars tell about the history of books, and he employed the word "sociology" in order to further expand the field's purview to include human agency.[19] The brilliance of McKenzie's approach was that even as it extended the scope of an already-overextended field or "tropical rain forest," it offered a compelling framework for revisiting old books and for imagining new ways of examining them. McKenzie claimed that a sociological approach to texts could lead to new insights and discoveries because "it can, in short, show the human presence in any recorded text."[20] And yet, as Mark Vareschi has argued, the groundbreaking work of book historians such as McKenzie "inevitably [has] to locate action and intention in known human actors" such that we

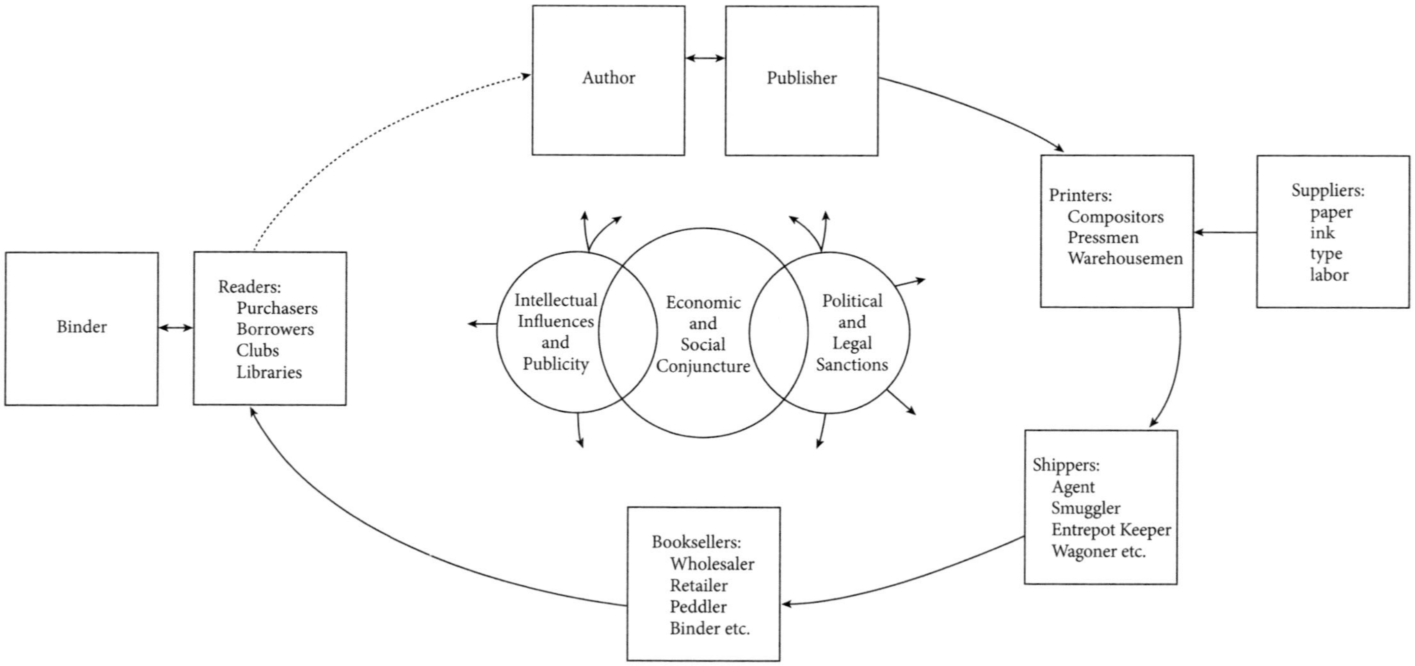

FIGURE 1. "The Communications Circuit," in Robert Darnton, "What Is the History of Books?" (1982). Reprinted courtesy of the MIT Press.

are left with "a delicate, and often unsatisfactory, balance being struck between the interpretive richness offered by their attention to the material text and the desire to justify such interpretations through recourse to the choices of authors, printers, publishers, apprentices, and others."[21]

What about nonhuman presence in texts, then? *The Nature of the Page* advocates an "ecology of texts" as a necessary and timely extension of McKenzie's sociology of texts.[22] Like a sociology of texts, an ecology of texts further extends the field, bringing more agents into play. It suggests, in a manner akin to that of McKenzie, that there is more to see and discover in the pages we have turned. An ecology of texts is an extension that has been latent in book history at least since Lucien Febvre and Henri-Jean Martin's seminal study, *The Coming of the Book*.[23] Febvre and Martin, like other scholars of the Annales school in 1960s France, are often credited with drawing attention to "the general pattern of book production and consumption" rather than "fine points of bibliography."[24] But in their discussion of papermaking as a collaboration of the "industrial" and the "natural," one can also see the seed of a more ecologically attentive approach: "The story, in brief, of the papermaking industry is that its development was always conditional on the supply of its raw materials."[25] Even the language of Darnton's communications circuit draws on ecological metaphors: "Books belong to circuits of communication that operate in consistent patterns. . . . By *unearthing* those circuits, historians can show that books do not merely recount history; they make it" (emphasis mine).[26] "Unearthing," though, requires one more step: the representation of the natural matter that makes the books that make history. One need not have any special investment in environmentalism to appreciate how much of the story we miss if we accept natural resources as givens rather than variables in the history of media.

On the other hand, one need not have any special investment in Renaissance poetry or book history to notice an important connection between the long, pollutive history of paper mills and Rob Nixon's attention, in *Slow Violence and the Environmentalism of the Poor*, to "paper trail identity" and the ways "writing has become fundamental to petro-modernity's control of labor."[27] Indeed, in the section "Orality, Geology, and Writing: The Technologies of Encounter" Nixon puts that work in a direct conversation with "The Sociology of a Text: Oral Culture, Literacy, and Print in Early New Zealand," the final chapter of McKenzie's *Bibliography and the Sociology of Texts*—and further highlights the "slow violence" that becomes visible when we attend to human presence in texts.[28] Early handmade paper, which has played such an important role in codicological and bibliographical research of the past, is filled with other kinds of legible evidence about histories of weather and weaving, of ports and imports and exports, of environmental exploitation and toxic effluence. As Diane Kelsey

McColley claims in a reading of Vaughan's poetry, "Part of ecological thinking is knowing where our artifacts come from, with what cost to the earth, to habitats, to species, to individuals of those species, and in human labor."[29]

Regardless of what we emphasize individually when we write about literature of the past, the great majority of humanities scholars access primary sources not only in editions, but also in archival libraries, where we hold and read and handle the slowly decaying remains of historical ecosystems that we call "Renaissance England" or "postcolonial Africa" or "nineteenth-century Latin American literature." (Throughout this book, I use the term "archival libraries" when designating specific, material spaces that facilitate on-site access to published and unpublished historical records.)[30] Bruce Holsinger suggests that "to be a medievalist is to be hopelessly implicated in and to constantly witness the mass deaths of countless sheep, lambs, calves, and goats for the means of literary transmission."[31] In doing so, Holsinger raises important concerns about the implications and the social and environmental ethics of archival research, topics to which I return in this book, especially in the final chapter. My starting point, though, is aesthetics as much as ethics, scholarly due diligence as much as environmental awareness. Seeing and emphasizing the organic nature of the texts we examine can give rhetorical weight to the matter from which texts are made and can begin to account for the crucial negotiations between humans and the world of things in the acts of creating (and preserving) material records of ideas.

We accept, as readers and as writers, that words on matter never perfectly replicate ideas in the mind. We accept and even enjoy these incongruities between mind and matter, but our fascination with the agency of humans, with human language, with human media, tends to blur our ability to see the ecological negotiations that undergird textual creation and circulation. Using biological metaphors like "medial ecologies," we tend to overlook the actual ecologies that make media possible.[32] And yet, as this book demonstrates, the affordances of a particular media form are determined by the affordances of the available plants, animals, and minerals used to create those forms. Natural resources can be imagined as boundless in Utopia; anywhere else, as the inputs from the natural world change, so do the outputs—the books themselves. A poem printed on paper can be read and circulated and preserved differently than a poem written on parchment or engraved on a stone monument or, for that matter, a poem inscribed on a cucumber.[33]

Confluence: Academic, Conservatorial, Artisanal

Inscribed cucumbers, as it happens, are a part of paper's fascinating history of raw materials to rags to reading matter, a history that has been told and retold in prose and in poetry for centuries. The story of paper is captivating in part because paper is both perfectly familiar, a quotidian object used in unremarkable ways each day, and perfectly exceptional, a remarkably important cultural commodity with an intriguing backstory. In *The New Organon* (1620), Francis Bacon says paper is both "a common enough thing" and "a *unique instance* of art," an "absolutely unique" artificial material that, although made from nature "change[s] her completely."[34] Fascination with the simultaneous oddity and mundanity of paper has inspired a vast body of literature on its history and uses, but the works that this book engages can be grouped into three kinds of expertise on early handmade paper, three streams of interest: academic, conservatorial, and artisanal. These streams of interest begin with different kinds of questions, are driven by different currents, and, as one would expect, end by telling stories in which certain kinds of details tend to surface. When they converge in *The Nature of the Page*, the currents shift and roil at the confluence, different kinds of details find their way to the surface, and a new, ecological story can be told about hundreds of years of human ideas entangled in plant fibers.

Academic studies focused on particular periods and/or techniques, especially those focused on sixteenth- and seventeenth-century handmade paper in England, are one set of sources that have contributed much to our knowledge of papermaking and paper use in recent years.[35] Despite its modest title, John Bidwell's essay "French Paper in English Books" still serves as one of the best, most concise introductions to papermaking in and around England from the mid-1500s through the late 1600s, and Helen Smith's recent essay, subtitled "The Proliferating Surfaces of Early Modern Paper," offers an overview of paper importation and papermaking in Renaissance England that brilliantly condenses the story into a dozen impressively documented paragraphs.[36] Period-focused introductions to the study of book history are similarly insightful as guides to early paper production and use.[37] Renewed interest in the materiality of textual forms has inspired a number of more focused studies that build upon standard works in the field, such as D. C. Coleman's *The British Paper Industry, 1495–1860* and Richard L. Hills's *Papermaking in Britain, 1488–1988*, while offering exciting new insights into the social circulation of paper.[38] Heather Wolfe's groundbreaking research on paper costs, which calls into question the often-repeated claim that writing paper was an expensive commodity in the Renaissance, stands out as an example of new research that corrects overgeneralizations that have long been accepted as fact.[39] Finally, because of its unique focus on pre-Gutenberg paper outside of Europe, Jonathan M. Bloom's *Paper Before Print: The History and Impact of Paper in the Islamic World* has proven an indispensable resource, one that strips away assumptions and reminds readers who typically handle sixteenth-century English books that the ink-imprinted, flax-based paper in those volumes is just one realized possibility, one version of paper among many.[40] Taken together as a stream of interest in paper and papermaking, these scholarly studies brim with questions about how paper came to be and how it came to be used, and their answers to these questions contribute invaluable historical context to the story that this book tells.[41]

Conservatorial interest in stabilizing and preserving the paper that has survived to the present day contributes another stream of vital information about paper and papermaking. "Because so much of our heritage . . . is paper-based, the importance of these artifacts is undeniable, and the need to conserve them is critical," writes Timothy P. Whalen, director of the Getty Conservation Institute, in a foreword to *Historical Perspectives in the Conservation of Works of Art on Paper*.[42] The volume, edited by Margaret Holben Ellis and comprised of a variety of historical perspectives on paper conservation, brings together centuries of dialogue about paper as a valued but imperfectly preserved object. A more focused but equally essential work in this category, John Krill's *English Artists' Paper: Renaissance to Regency* offers a fine-grained account of "the more than supportive role which paper has played in the graphic arts" told by an expert paper conservator with decades of hands-on experience in world-class museums.[43] More recently, *Fabriano: City of Medieval and Renaissance Papermaking*, by Sylvia Rodgers Albro, a senior paper conservator at the Library of Congress, offers an impressive, capacious history of papermaking materials and methods while focusing the narrative on a single locale.[44] Both Krill's and Albro's volumes are heavily and beautifully illustrated so that readers can see, for instance, photomicrographs of flax and cotton fibers at 50x magnification.[45] A significant development in the spread of conservatorial expertise between Krill's book (2002) and Albro's book (2016) is that in recent years, conservators at many institutions have begun publishing short, heavily illustrated, blog-style explorations filled with detailed, useful information about paper preservation. Recent posts to the British Library's *Collection Care* blog have included discussions of insect-damaged pages bound in manuscript fragments as well as "a leafcasting machine [that] can transform a damaged object in an instant with the help of paper pulp, gravity, and suction."[46] The conservation process, then, has grown increasingly visual and accessible as archives and libraries self-publish content on their websites and share images from the conservation lab on social media.

The artisanal work of hand papermakers contributes a third stream of expert information on early handmade paper. Broadly speaking, while book historians focus on recovering what was and while conservators focus on preserving what is, hand papermakers tend to be uniquely interested in imagining what can be. They approach the materials differently, ask different kinds of questions, and emphasize aspects of the process that a book historian might easily overlook. Like conservators, they think about fibers and fiber structure. They also think not only about the fibers that *have* been used historically for paper, but also about fibers that *can* be used—and the chemical composition of those fibers. For instance, in a brief introduction to the craft, papermaker Walter Hamady is quick to point out that "the main ingredient for all natural paper is simply CELLULOSE FIBER. Most living plants are made up of this fiber and, properly prepared, can produce some kind of paper."[47] Some papermakers, like Timothy Barrett, are actively striving to understand and replicate the techniques of early hand papermakers and publishing what they learn from that quest.[48] Barrett's website, Paper Through Time: Nondestructive Analysis of 14th- Through 19th-Century Papers, represents the gathered expertise of a master papermaker who is also deeply invested in conservation, and the introduction offered on that site, "European Papermaking Techniques, 1300–1800," is an essential contribution to scholarship on the history of papermaking.[49] The best broad overview in this category is Therese Weber's *The Language of Paper: A History of 2000 Years*, a comprehensive history of papermaking filled with details and insights and connections I have not found in other histories of paper.[50] Throughout this work I return again and again to the published work of papermakers like Hamady, Barrett, and Weber, to short articles printed in the journal *Hand Papermaking*, and to personal notes I have taken while interviewing master papermakers, particularly Robert Possehl.

Overview

The Nature of the Page is organized in two parts, "Legible Ecologies" and "Indistinct Ecologies," both of which emphasize the varied ways that humans recognize or do not recognize the plants, animals, and minerals in their media. Part I, "Legible Ecologies," takes up a narrative that is familiar to anyone who has engaged with the history of papermaking: handmade paper is made from other substances that circulated in societies, and not only papermakers and bookmakers but also readers and writers thought about the circulation of these materials. However, while previous studies of papermaking have tended to focus on the circulation of human commodities, especially rags, this work traces human commodities back to nonhuman materials such as flax, hemp, and straw. Part II, "Indistinct Ecologies," takes up interactions with handmade paper that have deep and abiding significance for writers and readers from past to present, but that have gone largely unnoticed and unread in the stories that we tell about books. What I show in these chapters is that when we pay attention to the ecology of texts, that is, when we question the broader range of human and nonhuman interactions that allow books to come into being, we begin to understand not just how humans use natural resources to convey ideas, but also how these nonhuman elements alter the ways human ideas can be expressed and even sometimes conceived.

The chapters in Part I, "Legible Ecologies," focus primarily on how writers and readers and printers used or negotiated with available ecologies in order to produce usable texts. In Part I, playfulness is an important part of the conversation, particularly the ways in which legibilities overlap and allow more poetic, playful, or surprising readings. In my use of the word, legibility—from the Latin for "readable" (*legibilis*), which is from the Latin *legere*, "to read"—can denote the ability to read an entire text and its component parts, not just the words on a page. Legibility, in this sense, allows a user to understand not just what can be done with the ideas conveyed by the materials, but also what can be done with the material conveyance itself (and its component parts). In Bradin Cormack and Carla Mazzio's *Book Use, Book Theory: 1500–1700* and in William H. Sherman's *Used Books: Marking Readers in Renaissance England*, the authors justify the use of the word "use" (as opposed to "reading") by citing an emblem in Geffrey Whitney's *Choice of Emblemes* (1586) with the motto *Uses libri, non lectio prudentes facit*, "Using a book, not reading it, makes us wise."[51] Book use as defined against some ideal of book reading has been productively explored in other periods in works such as Leah Price's *How to Do Things with Books in Victorian Britain* and in Rita Felski's *Uses of Literature*.[52] And Jeffrey Todd Knight has argued that "conceptions of utility . . . remain largely bound to modern categories, making room enough for *texts* but little for *textual objects*."[53] In response, Knight's work broadens the conversation about book use, considering, for example, books that were used primarily for furniture, not reading. More recently, Helen Smith has explored "the multitudinous uses of paper in early modern England," drawing special attention to the many uses of paper "that have nothing to do with writing."[54] And digital projects such as Cooking in the Archives and the Early Modern Recipes Online Collective are revealing, among many other things, just how important a role recycled paper played in Renaissance kitchens.[55] Legibility as use also productively contributes to conversations about historical literacy rates, since there may be a number of different kinds of literacies or use skills at play in regard to a work of literature.

Chapter 1, "Substances Used to Convey Ideas: Ship Sails, Cellulose, and Spinning Wheels," describes the transformation of a plant into a page of paper and considers how ecological scarcity has affected the history of the book from the beginning of papermaking in England until the latter half of the eighteenth century. It is easy to overlook the fact that most of the papermaking process involves working with materials that are not paper. Papermakers pulp and hydrate and draw out and dehydrate fibrous, absorbent plant matter as though they were alchemists transmuting a worthless substance—an old ship sail or a ragged smock, for instance—into an object as valuable as gold in terms of social history. (By contrast, parchment, an animal skin that has been prepared, not made, remains recognizable as an alteration of the thing it was before: a skin.) We know that paper's essential substance must derive from somewhere, and the first chapter demonstrates the importance of identifying what paper is and where it comes from. Placing paper history and, by extension, book history into a real ecological context rather than imagining it in a utopian nowhere, I argue that the standard narrative of rag shortages and eventual human innovation in the form of wood-pulp paper simplifies book history at the expense of environmental history. Despite centuries of international rag shortage and despite an abundance of local plants with suitable cellulose content for papermaking, hundreds of years pass without notable adjustments to well-documented ecological scarcity: Henry David Thoreau's paper is made from recycled rags just like Jane Austen's paper and Queen Elizabeth I's paper and Geoffrey Chaucer's paper.

Chapter 2, "The Word Made Flax: Cheap Bibles, Textual Corruption, and the Poetics of Paper," examines Henry Vaughan's poem "The Book" as part of a broader conversation about the poetics of paper: the rhetorical effects of the varied colors and qualities of paper used in the production of the vernacular Bibles that transformed reading practices in Renaissance England. The chapter also prompts a broader discussion about the rhetorical effects of flora and fauna in books, especially during the roughly four centuries (1450s to 1850s) when books were printed on paper made from recycled clothing. I find, in Vaughan's poem, a model of a Renaissance reader poring over an old book (a family Bible), pondering the legibility of its ecological history, offering hard evidence about the volume's structure. Like so many literary historians, Vaughan cannot resist the urge to narrate more than the evidence indicates. The history becomes an imagined story about the people who wore the linen clothing that became the paper, about the calf and the tree that became the book's binding, and about their future fate. The poetics of paper, the awareness that paper quality and prosody are intertwined, is rooted in an ecological mode of thinking, one that revels in the interconnectedness of heavenly ideas and earthly things sewn together in the form of a printed Bible.

The last three chapters, grouped in Part II, "Indistinct Ecologies," explore ecological aspects of books and paper that are easy to overlook, but that have shaped the ways humans used nonhuman materials to record ideas. From a subjective perspective, we might say they are illegible, but objectively speaking, the ecologies themselves are legible—perhaps unknown, but certainly, as I show, guessable. While I demonstrate that to read with a greater ecological awareness is to approach texts more like Renaissance readers, I am equally interested in what was missed, and even in what we can see now that they could not. Here Heidi Brayman's designation "abecedarian literacy" is useful, for though she uses the term in reference to "the most elementary vernacular reader" in the early modern period, we might think of ourselves as abecedarian readers of textual ecology.[56] While Chapters 1 and 2 consider ecology that was visible and notable to Renaissance readers and writers, Chapters 3, 4, and 5 examine aspects of book ecology that, while influential and important in terms of book production, were less visible to the average sixteenth- and seventeenth-century book user and that are nearly invisible to us now. Intrigued by these moments when things on which words are affixed become or are treated as invisible, illegible, indistinct, I work to fill gaps in our scholarly knowledge, but I also expose gaps and raise questions. Where what can be guessed is limited by available data, I try to shine light rather than shy away because I mean for this work to instigate more active attention to the ecologies from which our texts are made, and to the ways in which these ecologies inflect our engagement with the past and the future.

The title of Part II draws from and converses with *The Indistinct Human in Renaissance Literature*, particularly Jean Feerick and Vin Nardizzi's introduction to the collection in which they rely on the influential sixteenth-century theologian Richard Hooker's *Of the Laws of Ecclesiastical Polity* to call human exceptionalism into question: "In Hooker's account, a fully ripened man emerges as somewhat stunted in growth and capacity when positioned amidst the universe of other created things." Hooker, they observe, calls attention to "humankind's complex embeddedness among creaturely life on earth," to what the authors call "the potential for human indistinction."[57] The work of ecocriticism and the environmental humanities is filled with decentering metaphors such as "entanglement," "embeddedness," "networks"; however, "indistinct" calls attention to an observer who, making one thing come distinctly into focus, risks making other things indistinct. We might say we have to choose forest or trees, a metaphor that applies, but also one that brings us back around to paper, to the reminder that the ecologies of even this paragraph have been indistinct: Feerick and Nardizzi and the twentieth-century edition of Hooker that they cite are not persons, but trees. Their ideas came to me not in human-, but in tree-form. Typing these lines onto a paper-shaped rectangle on my computer screen, I expect to be similarly translated into words printed on tree pulp. Then the ecologies of these lines, of the many screens and scraps of paper used in rendering them readable in the form of this book or ebook, may be indistinct, but like Leopold's driftwood board, they will remain guessable.

Chapter 3, "How to Read a Blot: Historiography and Renaissance Ecologies of Inscriptive Error," opens by pondering a large blot on a single copy of Shakespeare's *2 Henry IV* in one of the Folger Shakespeare Library's First Folios. In that chapter, I show how one might read an indistinct stain on a page, not only metaphorically, but also materially, to better understand the mixed-media surfaces on which readers and writers (and printers) recorded and revised history from the time of Richard II (1367–1400), Henry IV (1367–1413), and Henry V (1386–1422) to the time of *Richard II*, *1* and *2 Henry IV*, and *Henry V*. Chapter 3 further demonstrates how the language of blots seeps into the King James Bible, marking another well-known king, King David, with an indelible reminder of his past mistakes. Ultimately, I argue that metaphors for thinking about error and reformation are linked to common writing technologies, specifically to the fundamentally different affordances of parchment and paper.

Chapter 4, "Sizing Matters: Annotating Animals in Renaissance England," reveals a surprising textual interaction between animals and readers. The most obscure natural ingredient in early printed books, and one that is almost uniformly overlooked despite scholarly interest in Renaissance reading practices, is "animal sizing," a viscous mixture made from gelatinized animal parts and used to coat the surface of paper so that ink would not blot into its fibers. Animals were integral elements of the literary ecosystem. Demonstrating how Renaissance reading habits and reading habitats are systemically linked, my work breaks new ground by arguing that scholarly data on book use and book survival in the period are skewed—and that future studies can produce more reliable data by recognizing the role of animals in paper books.

Focusing on John Donne's "A Valediction: Of the Book" and Shakespeare's *Sonnets*, Chapter 5, "This Book, as Long Lived as the Elements: Climate Control, Biodeterioration, and the Poetics of Decay," explores book biodeterioration across time by considering how Renaissance writers imagined book decay then, and how Renaissance scholars imagine book decay now. Taking its cue from the *Sonnets* themselves, the chapter explores the husbandry of paper memorials—how writers imagined paper as a living and dying monument that could outlast its own unique copy text. I explore these concerns, penned in the late sixteenth and early seventeenth centuries, from the point of view of a reader in an archival library in the twenty-first century, a century that if current environmental models prove true, could be the last full century in which original copies of Donne's and Shakespeare's poetry on paper can be preserved intact. The indistinct ecologies under consideration in this final chapter are the fossil fuels and nuclear energy used to create climate-controlled archival microbiomes where we can lose sight of the corruptible nature of books and of the poetics of decay.

Each of the five chapters explores a mode of negotiation with paper media: making paper (Chapter 1), reading paper (Chapter 2), editing on paper (Chapter 3), annotating printed paper (Chapter 4), preserving paper records (Chapter 5). And yet, focused as the chapters are on legibility and use—on readers and writers negotiating with an environment in order to record and transmit and store ideas—the chapter progression also works out the life arc of a paper text; this book's two parts, "Legible Ecologies" and "Indistinct Ecologies," describe the two broad phases in the ecological life of a handmade page: construction and destruction. Chapters 1 and 2 discuss details like cellulose and flax to show how a sheet of paper comes into existence in Renaissance England. Chapters 3–5 consider paper after its construction and examine how annotation and revision and nonhuman biodeterioration affect its contents and its uses (and users). But this work also suggests that we might pull the frame back even farther, revealing that the progression described above is not only about the life of a book, but also, and more fundamentally, about the lives of plants and animals and minerals as they move through society in textual form. Early handmade paper is a case study for a much broader methodology; I strive, in these pages, to use paper's ephemeral form to call attention to the ecosystemic relationship that always exists between human ideas and nonhuman matter.

Always our readings of texts are broadened or narrowed by what we acknowledge as part of the creation process. When we consider how the human species seeks to make sense of the world by using the organisms around them to make records of the past, when we consider how many plants and animals and minerals are housed in archival libraries and how vital these records are and may be to humanities and environmental humanities scholars alike, we could call disinterest in the ecology of texts an embarrassing oversight. We could also call it a prelude to a singularly exciting future direction for collaborative scholarship. Whether or not there is an ethical obligation to, in McKenzie's words, "show the human presence" in texts, we have long acknowledged that we cannot really make sense of texts in the world without accounting for human presence. And whether or not there is an ethical obligation to show the *nonhuman* presence in texts, I do not think we can make sense of the textual habits of humans without accounting for textual habitats, how the human species seeks to make sense of the world by using the organisms around them to make records of the past. Natural resources are provisional, seasonal, and geographically specific. To accept them as givens in the stories we tell about texts is to miss out on a vibrant history of the ecological negotiations and technological contrivances used to hold and share and save written records of human ideas.

PART I

LEGIBLE ECOLOGIES

CHAPTER 1

Substances Used to Convey Ideas: Ship Sails, Cellulose, and Spinning Wheels

> The art of Paper-making ought to be regarded as one of the most useful which has ever been invented in any age or country; for it is manifest, that every other discovery must have continued useless to society, if it could not have been disseminated.
>
> —Matthias Koops, *Historical Account of the Substances Which Have Been Used to Describe Events, and to Convey Ideas, from the Earliest Date, to the Invention of Paper*

> The *lack* of paper in England, due at first, no doubt, to the absence of any significant linen industry, meant that the Shakespearean text (like the vast majority of other English Renaissance texts) was a "foreign" body.
>
> —Margreta de Grazia and Peter Stallybrass, "The Materiality of the Shakespearean Text"

In John Gast's 1872 painting *American Progress*, an allegorical representation of America (as Columbia) leads pioneer expansion westward (see Figure 2).[1] Hovering above a river, she holds a large volume labeled "School Book" in her right hand, pressed against her breast; below the book and coiled around the crook of her right arm is a telegraph wire that strings along behind her.[2] In the sunlit landscape at Columbia's back, expectant trains cross open prairies westward toward cleared but train-trackless spaces. One of the trains appears to be set on a collision course with an enclave of teepees. Farther back and eastward, ships of different sizes sail around the tip of Manhattan and into the Hudson River, and the scene fades off in the upper right of the painting into rolling hills against

FIGURE 2. *American Progress*, chromolithograph by George A. Crofutt (ca. 1873) after 1872 painting by John Gast.

a bright morning sky. It is the East, and Columbia is the sun. Before her, a shadowy landscape awaits refinement. Deeper into these shadows, Native Americans and buffalo and bear and deer flee. We cannot see the Pacific Ocean, a boundary that might suggest the limits of expansion.[3]

Like the narrative depicted in *American Progress*, the narratives we tend to tell about the history of paper advance from past toward present, from East toward West, and from human ingenuity toward nonhuman, natural resources. In Dard Hunter's seminal *Papermaking: The History and Technique of an Ancient Craft* and in any number of scholarly and/or popular histories of papermaking, we find a chronological narrative that pans from evidence of the invention of papermaking about two thousand years ago in China (in the first or second century BCE) toward its spread across the world.[4] By the end of the eighth century BCE, papermaking had spread to Samarkand (present-day Uzbekistan) and then on to

Baghdad. By the thirteenth century, paper was being made in Egypt, Morocco, Syria, Spain, France, and Italy. By the end of the fifteenth century, it was being made in Germany and England. In 1690, the first paper mill, Rittenhouse Mill in Pennsylvania, was set up in the United States. By 1854, the first paper mill west of the Missouri River began operating in Salt Lake City, Utah.[5]

In these past-to-present, East-to-West accounts, a human desire to produce a media that can be used for recording takes center stage. Religious zeal, for instance, is often credited as the impetus for the spread of papermaking: Buddhist missionaries first bring papermaking technology from China to Japan and Korea; Muslims bring papermaking to Iraq, Syria, Egypt, North Africa, and, eventually, to Italy and Spain; Christian crusaders bring papermaking even farther west; by 1860, Brigham Young starts a paper mill in Utah.[6] Nonhuman materials are discussed at critical transition points in the narrative—from plant-sourced papyrus to animal-sourced parchment to plant-sourced paper, from mulberry fibers in the East to linen rags to tree pulp in the West—but then they tend to disappear from view even as the very technology that brought those nonhuman materials into view begins to overwhelm and exhaust natural supply. But when, inevitably, resources become scarce, the nonhuman reemerges in the papermaking narrative as human innovators ask: What other substances might we be able to record ideas upon? This linear narrative from past, East, and human toward present, West, and nonhuman is an effective and compelling way of organizing a story that continues to be of interest to specialized and general audiences alike.[7] It is also, like every story—like this story or like the one depicted in *American Progress*—a structure that makes some features more prominent and more important and other features less noteworthy and less meaningful. This chapter reverses the usual narrative, opting instead to tell the history of paper from present toward past, from West toward East, from nonhuman to human, offering a perspectival shift that allows us to see a plant-based media that has changed human and environmental history. It is a perspective, too, that allows us to see how little papermaking technology changed in Europe from the late 1300s to the mid-1800s. Only in the last century and a half has the papermaking industry relied on trees. For most of the history of printed texts, plant fibers, worn and used to rags by humans, then recycled, have been the primary source of material for papermaking.

Legible Sails

Ships, like those whose bright sails catch our attention on the eastern seaboard of Gast's *American Progress*, help to frame this chapter about the plant substances that are used to convey ideas. This chapter's reverse-chronological

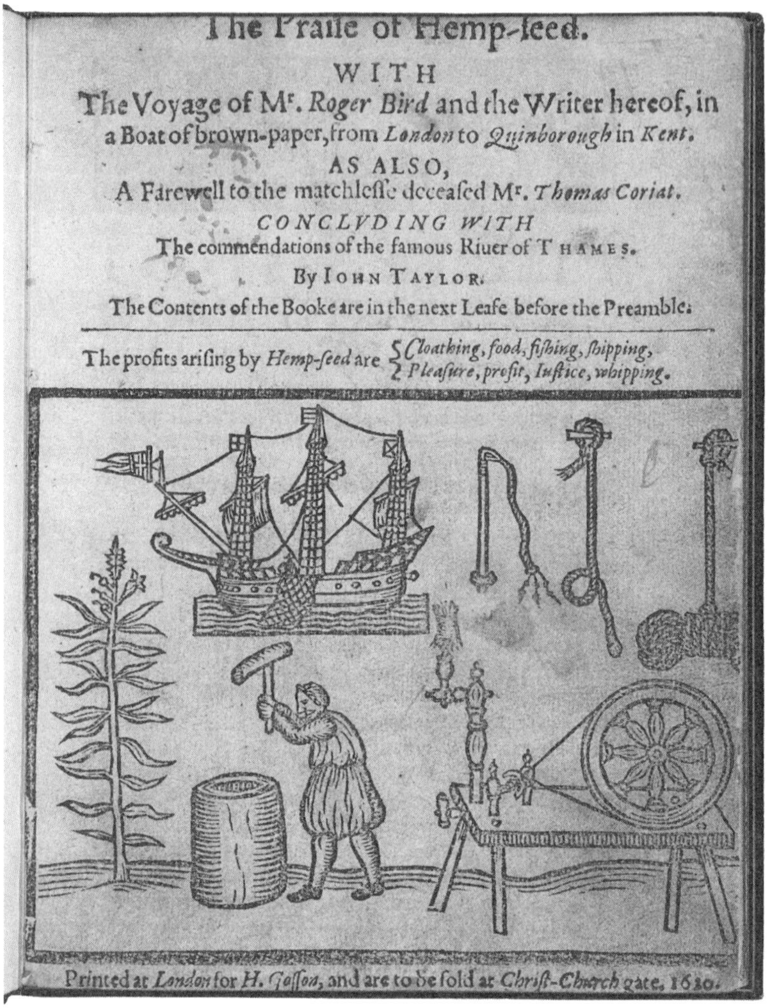
The Praise of Hemp-seed.

WITH

The Voyage of Mr. *Roger Bird* and the Writer hereof, in a Boat of brown-paper, from *London* to *Quinborough* in *Kent*.

AS ALSO,

A Farewell to the matchlesse deceased Mr. *Thomas Coriat*.

CONCLVDING WITH

The commendations of the famous Riuer of THAMES.

By IOHN TAYLOR.

The Contents of the Booke are in the next Leafe before the Preamble.

The profits arising by *Hemp-seed* are { *Cloathing, food, fishing, shipping, Pleasure, profit, Iustice, whipping.*

Printed at *London* for *H. Gosson*, and are to be sold at *Christ-Church* gate. 1620.

FIGURE 3. Title page from John Taylor, *The Praise of Hemp-seed* (1620), sig. E4r. Used by permission of the Folger Shakespeare Library.

narrative begins with Henry David Thoreau watching tattered ship sails pass by Walden Pond in America ca. 1840 and ends with John Taylor, "The Water Poet," in "a Boat of brown-paper" that is sinking into the Thames River in England ca. 1620 (see Figure 3).[8] Ships, especially ship sails, are a shared point of reference for both writers. If we imagine some trick of spacetime by which these two writers find themselves side by side in an art gallery examining *American Progress*, we realize that Taylor would be confounded by the telegraph wire that strings all the way back to the Atlantic Ocean, and Thoreau would need to explain that technological innovation to him. However, the masted ships sailing in New York Bay and the book in Columbia's arm would have been familiar enough to both writers. Both would have understood the ecosystemic relationship between ship sails and books, an ecosystemic relationship that was crucial to human communications from the spread of papermaking technology in medieval Europe through the Industrial Revolution and right up until the end of the nineteenth century.

Taylor would also need Thoreau to explain the locomotives puffing their way westward, and we can guess from chapter 4 of *Walden* that Thoreau might do so with conviction. Few sections of *Walden* mar the notion of Thoreau as a solitary hermit writing in the wilderness quite so much as the chapter "Sounds," in which Thoreau devotes his attention not to the noises and songs of the pond's inhabitants, but to the sounds and rhythms of the railroad just south of the pond. Two chapters earlier, Thoreau leveled an oft-quoted critique of the railroad—"We do not ride on the railroad; it rides upon us"—but in his chapter on "Sounds," he likens the train's whistle to "the scream of a hawk" and expresses appreciation for its daily rumblings: "I am refreshed and expanded when the freight train rattles past me, and I smell the stores which go dispensing their odors all the way from Long Wharf to Lake Champlain, reminding me of foreign parts, of coral reefs, and Indian oceans, and tropical climes, and the extent of the globe."[9] Thoreau writes about brandy headed north to farmers in the Green Mountains, and about cattle cars headed south out of those farms. He looks up from his book and sees "some tall pine, hewn on far northern hills" on its way to becoming a ship mast.[10]

As Thoreau may have known, the pine-trees-as-ship-masts conceit recalls the Old English poem "The Dream of the Rood," a prosopopoeic narrative told by the tree that became the cross on which Christ died. The traveling-tree conceit also feints toward the popular "it-narratives" of the late eighteenth and early nineteenth centuries.[11] In one relevant it-narrative that has received ample critical attention, "The Adventures of a Quire of Paper," a man reading a printed sermon in a coffeehouse is suddenly accosted by the paper in his hands. "Though now offered under this form to thy eyes," narrates the page, "yet know my original form was that of a *thistle* in the ditch that surrounded a farmer's meadow."[12] The thistle then relates a story of pride and glory and devastation, of being harvested with the flax in neighboring fields and of hardships on the way to becoming a printed sermon. We are not surprised, at the end, to learn from the human narrator that it was all a dream. Like the tree in "The Dream of the Rood," the talking thistle/quire of paper was "merely the vision of a drowsy fit."[13]

When, in the midst of his reflections on trains and sounds, Thoreau thinks about torn ship sails headed north to be recycled, he ventures even closer to an it-narrative like "The Adventures of a Quire of Paper": "This car-load of torn sails is more legible and interesting now than if [the sails] should be wrought into paper and printed books. Who can write so graphically the history of the storms they have weathered as these rents have done? They are proof-sheets which need no correction."[14] Pulp them into paper and print *Moby-Dick* on them and they will not tell better stories of the sea than the stories they already offer in their current, ragged state. And yet these sails are insistently non-prosopopoeic in Thoreau's telling. If, as Christina Lupton claims, "the defining feature" of it-narratives is that "they are told from the perspective of an object or (less commonly) an animal," then Thoreau's sails do not qualify, for though they have recorded the history of the storms, they do not speak up.[15] In "Rood" and "Paper," the dreaming readers or auditors are accosted by objects. In Thoreau, by contrast, a wide-awake reader reckons worn-out sails to be "more legible and interesting" than a sea narrative printed on paper. Legible to whom? Is Thoreau implying that, touching his fingers to the wounds in the sides of these sails, he can recount a detailed narrative of a storm at sea, something rivaling John Donne's "The Storm"? Perhaps there is a reader, some ancient mariner, to whom each stain and scar recounts a detailed story, but most would rather read *Sea Wolf* than a sail.

Watching the train pass and imagining its cargo, Thoreau moves from torn hempen sails and their relationship to books to worn-out cotton and flax rags and their relationship to real-life narratives: "These rags in bales, of all hues and qualities, the lowest condition to which cotton and linen descend, the final result of dress,—of patterns which are now no longer cried up, unless it be in Milwaukie, as those splendid articles, English, French, or American prints, ginghams, muslins, &c., gathered from all quarters both of fashion and poverty, going to become paper of one color or a few shades only, on which forsooth will be written tales of real life, high and low, and founded on fact!"[16] Within a couple of pages, the train has passed and Thoreau is back to listening to bird sounds, like the dismal scream of a screech owl, which, he claims, is "truly Ben Jonsonian."[17] In other words, soon after Thoreau ends his meditations upon the paper futures of sails and rags, he summons the ghosts of rags and sails past. Indeed, Thoreau scrawls all of these observations on the same kinds of material Jonson used to scrawl his ideas: flax, cotton, and hemp from recycled clothes and sails.

The sameness of source materials across centuries is most apparent in John Taylor's celebratory poem *The Praise of Hemp-seed* (1620) (see Figure 3). In *Hemp-seed*, Taylor observes the same relationship between sails/clothing and paper that Thoreau rhapsodizes about. The hemp plant, writes Taylor,

> makes clothes, cordage, halters, ropes and sailes.
> From this small Atome, mighty matters springs.[18]

No stranger to the prosopopoeic narrative (Taylor's 1630 collected works include *Hemp-seed* as well as an it-narrative called *The Trauels of a tweluepence*), in *Hemp-seed*, Taylor opts to tell the story as a wide-awake human narrator who celebrates hempseed's many uses and then celebrates its second use in the form of paper (more than half of the poem is devoted to praising hempseed-as-paper). Less than halfway through the poem, Taylor writes,

> But paper now's the subiect of my booke,
> And from whence paper it's beginning tooke:
> How that from little *Hemp* and flaxen seeds
> Ropes, halters, drapery, and our napery breeds,
> And from these things by Art and true endeuor,
> Al paper is deriued, whatsoeuer.[19]

What we see, then, is that when Thoreau writes about papermaking materials at Walden Pond in the 1840s and/or when revising his manuscript in the 1850s, he describes the same raw materials Taylor was writing about across the Atlantic Ocean nearly two and a half centuries earlier. The material substance of paper is never really invisible, but it is always assumed. For most of the history of the book, rags make paper. It becomes a truism, even a common way of conceptualizing historical communities.[20]

Reading and comparing these two passages from Thoreau and Taylor, written two centuries apart on two separate continents, we might mistakenly assume that papermaking from hemp-, cotton-, and flax-based rags was a remarkably stable, sustainable enterprise. And yet we know from historical accounts that between the time of Taylor and Thoreau, both England and America faced severe rag shortages. In *A History of Paper-Manufacturing in the United States, 1690–1916*, Lyman Horace Weeks claims that "the history of paper-making in Europe and in the United States is shot through and through with the records of persistent speculating and experimenting in the endeavor to escape from the limitation imposed upon it by sole dependence upon rags."[21] The young, independent American "Republic of Letters" depended heavily on European rags—rags that were already in short supply in Europe. By the end of the eighteenth century, newspaper advertisements routinely admonished households to recycle their rags for the good of the community and, eventually, for the good of the nation, rather than reusing rags within the home. Susan Strasser cites newspaper advertisements as "evidence of perpetual rag shortage in the colonies and the new nation and of the papermakers' strategies of propaganda, education, and entreaty, aimed at gathering enough rags to keep new mills running."[22]

It is shocking, in retrospect, to consider how long it took for wood pulp to replace flax (and cotton and hemp) as the primary raw material from which paper is made, especially when "for more than a hundred years, the existence of the industry was constantly imperiled by this scarcity."[23] Weeks writes of the "curious and exceedingly interesting chapter" in American history "which treats of the persistent and not always successful struggle for raw material to keep the mills going."[24] He goes on to cite lists of hundreds of raw materials from which experimental paper was made, lists that include substances such as animal excrement, asbestos, asparagus, bananas, beets, brewery refuse, corncobs, cucumbers, dust, hop vines, horseradish, moss, peat, pineapples, pine shavings, turnips, and seaweed.[25] American papermakers began experimenting with local natural resources as early as the 1790s as numerous organizations, including the American Philosophical Society and the American Company of Booksellers, offered prizes for innovators who could discover a suitable replacement for rags. Woven into the language of these inquiries and experiments is a patriotic optimism that the untapped landscape of America would yield a new plant-based substitute for rags. In 1785, the poet Philip Freneau reimagined Palemon, a character who appears in Chaucer's *The Knight's Tale* and Shakespeare's *Two Noble Kinsmen*, immigrating to America "to tame the soil, and plant the arts."[26] Well into the nineteenth century, American innovators were still struggling to find the plant that would help them escape dependency on European rags and realize their ambition of planting the arts. Back in Europe, some innovators were similarly drawn to the idea of making paper from native resources—not from an exotic new plant, but from some overlooked but viable source.

Fascinating stories of rag recycling, of rag-collecting propaganda, and of shocking materials like asparagus and excrement used to make paper are not uncommon. They tend to put all of the emphasis on humans, on book and paper consumers, and they often read like success stories where nations and industries, led by brilliant innovators and scientists, overcome austerity and adversity. As readers, we move through many of these bildungsroman-style narratives by turning paper pages, so we know there is a happy, paper-supplied ending. Maybe we even unwrap reams of printing paper emblazoned with arboreal logos so we can print out parts of these stories. Too often, though, the history of papermaking tacitly elides the fact that it took too many innovators far too long to find a suitable substitute for rags.

The Transformation of a Plant

In *Print, Manuscript, and the Search for Order, 1450–1830*, David McKitterick writes, "The real issues in papermaking lay in the discovery and development of raw materials to meet a growing demand from the printing industry. From the mid-[eighteenth]century onwards—and with mounting urgency as curiosity was fed by need—wood, straw, nettles and other vegetable matter were subjects of experiment and of competition, in a search for a material that was plentiful, cheap and reliable."[27] True as the observation is, it belies an intriguing conundrum: Which was the priority—"urgency" and "need" or vegetable matter that was both "cheap and reliable"? Historically, papermaking technology has been driven not as much by demand for paper as by demand for raw materials that could be obtained cheaply. The history of paper production, like the history of oil production, is a history of scarce nonhuman supply pitted against self-assured human demand. When "cheap and reliable" are the standards, especially where a sense of urgency is in play, a sustainable bargain, one that tends to balance supply and demand in a given ecosystem, is unlikely.

According to a well-known poem quoted by Dard Hunter in *Papermaking*,

> RAGS make paper,
> PAPER makes money,
> MONEY makes banks,
> BANKS make loans,
> LOANS make beggars,
> BEGGARS make
> RAGS.[28]

In these lines, as in Darnton's Communications Circuit (discussed in the Introduction), cycles of human use within a social economy are imagined as independent from the natural world. But rags do not simply come to be because of defaulted loans. To better understand rag shortage and the search for a suitable—that is, a cheap—plant substitute, we might tell a less anthropocentric narrative that begins well before the rag stage. "Papermaking is the transformation of a plant," claims papermaker Robert Possehl.[29] It is a concise definition that literally grounds the study of papermaking. Walter Hamady notes that "the main ingredient for all natural paper is simply CELLULOSE FIBER. Most living plants are made up of this fiber and, properly prepared, can produce some kind of paper."[30] So we might say:

> SEEDS make plants,
> PLANTS make cellulose,
> CELLULOSE makes textiles,
> TEXTILES make rags,
> RAGS make paper.

Often, the story of papermaking is a story of solutions, of what happens after the right plant has been "properly prepared," but here I wish to slow down and emphasize the problems of cellulose extraction because that part of the process can give us a better understanding of paper's varied ecologies across time and place.[31]

Cellulose does not often appear in book history scholarship, but it plays a central role in the story of paper's ecology. Cellulose, by definition, is "an insoluble carbohydrate which is the main structural constituent of the cell walls of plants, and one of the most abundant organic compounds on the earth."[32] Scholars and papermakers alike tend to refer to paper made from straw or linen rag or flax or wood, but those designations might be thought of as shorthand: paper is made from the cellulose present in each of those plants.[33] We use the same sort of shorthand when we refer to dietary fiber: we mean, primarily, cellulose, the part of a plant that is not broken down in the digestive tract. Because book history has tended to tell a more or less anthropocentric story about paper, cellulose rarely merits inclusion in the narrative.[34] Mark Bland's *A Guide to Early Printed Books and Manuscripts* and John Carter and Nicolas Barker's *ABC for Book Collectors* cite "cellulose" once each, for example. In both cases, the resilience of cellulose fibers in forming a durable book object is focal: Bland notes that the "cellulose fibres of paper withstand folding," and Carter and Barker state in their definition for "de-acidification" that "earlier paper made from rags or esparto grass contains pure uncompounded cellulose, and, short of physical assault, has a very long life."[35] Cellulose, in book history, plays an important, but limited and subdued role.

Turning from the scholarship of book history toward that of artisans, conservators, and scientists, we find cellulose to be the real hero of the story of papermaking. It actually does seem that dramatic at times—as if we have all been aware that Clark Kent plays a key role in the Superman narrative without guessing his true identity. In *The Complete Book of Papermaking*, paper artist Josep Asunción explains that "what we call paper is actually a thin sheet produced from the physical bonding of previously hydrated fibrous materials, mostly cellulose."[36] In the preface to the English edition of Therese Weber's *The Language of Paper*, Jonathan M. Bloom defines paper as "technically a mat of cellulose fibers suspended in water, deposited on a screen, and then dried," and throughout the book, Weber circles back to a dependency on "raw materials . . . found locally" and on human struggles to extract cellulose from those raw materials.[37] She calls attention, for instance, to the repeated failures of European papermakers to make paper directly from plants rather than from rags and claims that when late nineteenth-century researchers "succeeded in extracting wood cellulose by the sulphite or sulphate process, the use of wood fibre became properly practicable for the first time."[38] Richard L. Hills, whose *Papermaking in Britain, 1488–1988* focuses on the industry, technology, and science of papermaking, defines "cellulose" as "the basic substance of paper manufacture."[39]

Noting the different ways that academic and artisanal narratives are told, and especially the different beginning points of those narratives, I put what I intend to be productive pressure on the underlaps of academic and artisanal expertise, and I do so with the aim of calling attention to productive overlaps between book history and environmental humanities scholarship.[40] A key insight into the ecology of paper that arose from thinking with artisans who make paper by hand using historical techniques was the simple but profound realization that careful, lengthy prep work is key to a well-made final product, and perhaps the most important step in this prep work is the process of fiber maceration known as retting. In simple terms, retting is the process of stripping away all but cellulose. Numerous methods were used: papermaker Timothy Barrett writes that "the fermentation methods used by various mills are very likely to have differed as much as the construction and location of the mills themselves."[41] Barrett emphasizes the specialized skill that went into the essential process of retting, likening it to the work of a good winemaker: "Knowing how to ret rags was not unlike knowing how to ferment grape juice to make good wine." As with wine, Barrett notes, the methods used to ret paper yielded unique final products: "Retting is a crucial reason for the unique look, feel, and handle of many of the best early book papers . . . second in importance only to the special nature of the old-rag raw material."[42] To make paper by hand, to try to craft even a mediocre replica of the kind of surface book historians encounter by the hundreds and thousands, is to begin to understand that papermaking begins many steps before the vatman dips a mould and deckle into the vat.

In the preface to *Papermaking by Hand*, Hamady recounts that "a scholar who had written a book on papermaking came to visit one time and when he saw the actual formation of the sheet, loudly exclaimed 'so that's how it's done!' "[43] I sympathize with the scholar in Hamady's anecdote; I only began to appreciate the crucial role of cellulose in the papermaking process while standing in the snow in Portage, Wisconsin, about eleven miles as the crow flies from where Aldo Leopold found the driftwood discussed in the Introduction. I was using a canoe paddle to stir a cauldron in which Robert Possehl was boiling down a herbaceous plant known locally as Lady's Mantle—identified in Renaissance herbals by the same name or, alternately, "great Sanicle" or "Lyons Foot."[44] We were cooking plants, not retting rags, but our object was the same. "Cellulose is what paper is made from," Possehl said, and the process of cooking is one way to reduce the plant to pure cellulose. I relate this story, anecdotal as it is, in an attempt to properly cite the sources from which I have borrowed insights. This is a book that could not have been written well by relying solely on other books; rather, it relies on the generosity and insight of scholars and artisans, many of whom, like Possehl, have not published the unique insights they have gained from decades of research and experimentation. Working, however, with someone who excels at making paper directly from plants, I began to understand how the need for more "cheap and reliable" sources of cellulose affected the history of the book in the first place.

Cheap cellulose would be cellulose that is both abundant *and* easily extractable. All plants contain cellulose, but in varying amounts. For instance, cellulose makes up "about 90 percent of cotton and about 50 percent of wood."[45] Fiber length matters, too. Flax has about 10 percent less cellulose than cotton, but "its fiber was longer and stronger, and its fiber wall was straighter and thicker."[46] In rag form, plant fibers came to papermakers pre-processed: the labor required to convert the plants into textiles had already been done. The labor of further breaking down the fibers through use (as clothing or ship sails, for example) processed the plant fibers even more. So no matter what the plant fiber, much less labor was required to pulp and make paper from linen rags or ship sails than directly from flax and hemp plants. We might visualize the rag shortage crisis as a triangle where the simplest solution would be to cut out the textile phase and make paper directly from plants (see Figure 4). Thus we see so many lists of experimental plants in the history of paper, and so many years passing without a viable paper product. Then Matthias Koops, an immigrant to England at the end of the eighteenth century, claimed the problem of scarce, expensive, imported rags could be solved with a readily available, cheap, local raw material: oat and wheat straw.

Substances Conveying Ideas

In autumn of 1800, an odd book printed on coarse yellow paper began circulating in London: *Historical Account of the Substances Which Have Been Used to Describe Events, and to Convey Ideas, From the Earliest Date, to the Invention of Paper* by Matthias Koops (see Figure 5). The title page advertised its contents as a history of media told as a history of raw materials from which media were made, a history not of great books and big human ideas, but of nonhuman "Substances" used "to Describe Events, and to Convey Ideas." The title page also advertised its own material history: "Printed on the First Useful Paper Manufactured Soley [*sic*] from Straw."[47] The vivid yellow hue of the paper is heavily flecked with organic matter. In archival libraries, under direct light, the paper can take on a shimmering quality as brownish flecks spark into gold

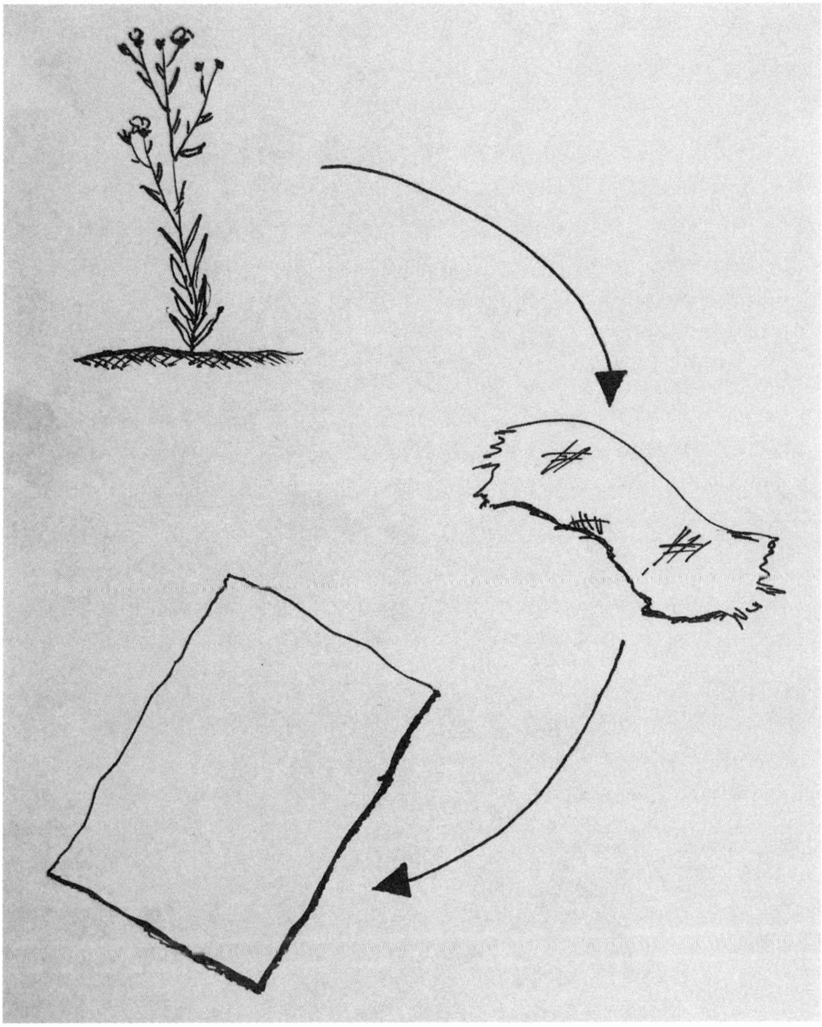

FIGURE 4. "Plants to Rags to Paper," Joshua Calhoun.

flashes that appear and disappear as the pages are turned. The pages are rough to the touch, and page thickness is uneven throughout. The author's claim that the paper is made only from straw is grander than it might sound, and it comes from someone who had a history of making grand claims that did not always prove true. And yet, more than two hundred years later, the paper used in the 1800 first edition and in the expanded 1801 second edition is consistently pliable and strong across numerous copies held in multiple locations.[48] Koops acknowledges in *Substances* that his innovation was part of a conversation that had been happening in Europe for nearly a century among figures like Carl Linnaeus,

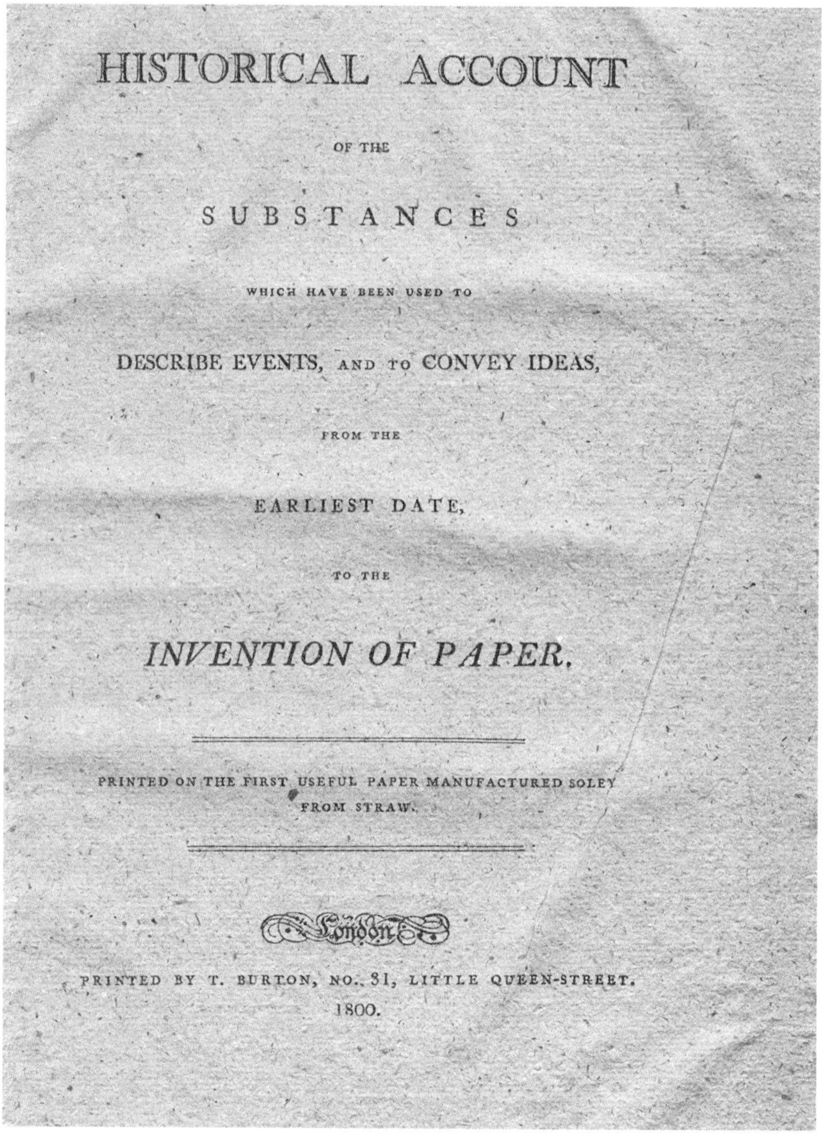

HISTORICAL ACCOUNT

OF THE

SUBSTANCES

WHICH HAVE BEEN USED TO

DESCRIBE EVENTS, AND TO CONVEY IDEAS,

FROM THE

EARLIEST DATE,

TO THE

INVENTION OF PAPER.

PRINTED ON THE FIRST USEFUL PAPER MANUFACTURED SOLEY FROM STRAW.

London:

PRINTED BY T. BURTON, NO. 31, LITTLE QUEEN-STREET.

1800.

FIGURE 5. Title page from Matthias Koops, *Historical Account of the Substances Which Have Been Used to Describe Events, and to Convey Ideas, From the Earliest Date, to the Invention of Paper* (1800), printed on straw paper. Courtesy of the Department of Special Collections, Memorial Library, University of Wisconsin–Madison.

René Antoine Ferchault de Réaumur, Jacob Christian Schäffer, and a dozen other "scientific men" who had recorded (on rag paper) their "ideas on substitutes for paper-materials."[49] Réaumur, observing wasps, realized that paper could be made from wood as early as 1719. Schäffer was making paper from wood and straw (as well as cattails, wasps' nests, hops, moss, and cabbage stalks, among others) in the 1760s.[50] "It is surprizing," writes Koops, "that the observations of [these] authors . . . should not have been earlier attended to by . . . intelligent paper-makers, who had the road thus opened to them for their investigation."[51] Up to this point in the book, Koops has made the point that over and over throughout recorded history, human ingenuity confronts a scarcity of nonhuman materials, always with the same outcome: an abundant substance is made into a recording surface that supersedes what came before it. Koops relies on the authority of scientists to argue that rag paper is ready to be superseded: "These authors have stated, that as cotton, flax, and hemp, are the origin of paper and rags, other vegetables of a tender and pliable nature might probably be converted into mucilaginous pulp, and adopted in the manufacture of Paper; and farther, that those vegetables that are of a brittle and harsh nature, but which can be obtained in large quantities and at moderate prices, might by art and perseverance be made tender, without destroying that quality which is necessary to be retained in paper-stuff."[52] Koops set out to make paper from straw not because he believed he could make better, whiter, more beautiful paper from that substance, but because he believed he could make perfectly serviceable paper from straw that was also, though he would not have used the phrase, ecologically sustainable.

Substances is perhaps the most stunning conjunction of form and content that I have held: a book about the history of books printed on mass-producible paper made by replacing scarce, imported rag fibers with locally grown, abundant oat and wheat straw. Koops's volume introduces itself as a history and a future of the book, and it outlines ways of thinking about book history that upend the most popular ways we have schematized the field of study. *Substances* is reference book and rare book, information bank and artifact. My home institution's Special Collections owns two copies: one is in the vault alongside rare books, and the other is in the open reference stacks in the Special Collections library right alongside other modern books on the history of papermaking. The Library of Congress Subject Heading (LCSH) "Papermaking—History" and "Writing materials and instruments" groups *Substances* with indispensable reference works on paper that I have already cited in this chapter, such as Dard Hunter's *Papermaking: The History and Technique of an Ancient Craft* (1947) and Therese Weber's *The Language of Paper: A History of 2000 Years* (2007). The paper in Koops is lousy by comparison to the paper in many of the fifteenth- and sixteenth-century holdings, but compared to its neighbors on the reference shelf in Madison, the paper is beautiful, sturdy, and resilient.

Koops's volume, with its titular emphasis on substances (not substrates), with its odd, coarse, yellow pages, inspires a different kind of story about the history of books and especially paper. *Substances* refuses to be a transparent medium made of limitless raw materials. On pages visibly made from straw, Koops narrates how various raw materials (trees, stones, metals, wax, bark, leaves, papyrus, skins) have been used historically "to preserve the remembrance of important events."[53] It is a history of things used to convey history. Negotiation among humans and nonhumans is at the heart of Koops's story; the plant sources never drop from view or fade into the background. Because Koops writes in the midst of serious material scarcity, because the stuff of paper cannot be assumed as it so often is in book history narratives, because Koops himself has a stake in making the paper he describes, he is attuned to its plant origins in ways that we tend to overlook. *Substances* is a remarkable, and remarkably understudied, resource that deserves renewed attention.

Substances reminds its reader, over and over again, that writing surfaces used as conveyances for human ideas were first nonhuman matter. Koops begins his history of substances used to convey ideas in the ancient past, when "trees were planted, heaps of stone, or unornamented altars and pillars, were erected . . . to keep up the recollection of past facts."[54] As writing and alphabet systems developed, more complex means were used to recollect the past: "Since the art of writing was invented, several materials have been used on which was engraved or written what was wished to be conveyed to posterity," and in short time, claims Koops, different kinds of substances were used to record different kinds of public and private writing: "A distinction has been made between public records and private writings. For the first; stones, timber, and metals, were chiefly used; and, for the latter, leaves and bark of trees."[55] The story that Koops tells may begin as a narrative of diversified formats used to distinguish between private and public records, but eventually, he settles into an arcing narrative in which one substance/substrate supersedes another substance/substrate. "The use of boards was *superseded* by the use of the leaves of palm, olive, poplar, and other trees," he writes; then "the custom of writing on leaves of trees was *superseded* by the use of the raw bark of trees, and the interior bark" (emphasis mine).[56] Pausing here, he notes that the language of books is rooted in trees: interior bark is called *liber* in Latin; these barks, rolled up so they could be carried about, are *volumen*; the word *codex* "notwithstanding its true meaning is the trunk of a tree . . . was adopted to describe many sheets of the said bark together."[57] Koops devotes some space to discussing various kinds of bark media from various locales, but only after noting that the use of inner bark (*liber*) is a short step away from making papyrus. In the narrative laid out in *Substances*, it becomes readily apparent that had northern-growing birch trees been more suitable for making writing substrates than southern-growing papyrus, the history of the book, and perhaps the history of civilization, would be quite different. Ecological availability and scarcity—we might say biodiversity—determine the course of paper and books in Koops's narrative.

Intriguingly, when Koops turns to a consideration of parchment superseding papyrus, he tells an often-repeated anecdote about Ptolemy/Eumenes, but in a way that emphasizes not only political power, but also ecological scarcity and fear.[58] In the usual story, Ptolemy wants a bigger, better library in Alexandria than Eumenes has in Pergamus, so he cuts off exports of papyrus (which Koops refers to as "Egyptian Paper," though he clearly indicates that it is a precursor to paper). However, Koops suggests that Ptolemy's decision might have been driven by ecological concerns: "It may be that this prohibition was not solely occasioned by jealousy, but from the fear that his dominions . . . would be again reduced to a state of ignorance for want of Paper, because the plant failed sometimes in unfavourable weather."[59] Having been cut off from the raw materials needed to record and convey ideas, the Pergamians, according to Koops, "were therefore obliged to devise other means for making Paper." In response, they invent parchment, a writing material that "obtained its name from the city of Pergam, or Pergamus . . . the place where it was invented."[60] The narrative of supersession continues: born out of necessity and ecological scarcity, parchment "greatly surpassed the Egyptian Paper in fineness, smoothness, and strength."[61] Parchment, in Koops's telling, also surpassed papyrus, because it was made from a reliable, sustainable, abundant, and geographically nonspecific material. The spread of parchment and especially parchment-making, he notes, was far broader than the spread of papyrus because, once learned, the techniques of parchment-making can be practiced in any climate where animals can be raised.

So, according to *Substances*, boards as a writing medium are superseded by leaves, which are superseded by the inner bark of trees, which are superseded by the rinds of papyrus plants, which are superseded by parchment. The title promises a historical account of substances "From the Earliest Date, to the Invention of Paper," and at the bottom of page 46, almost exactly halfway through the 91-page volume, Koops turns to paper. His history of paper and papermaking follows the same pattern as his history of substances used before the invention of paper: paper from linen rags superseded paper from cotton rags partially because flax grows more abundantly in Europe. Notably, Koops claims "that the Greeks made use of cotton-paper sooner than the Latins; and that it was brought into Europe by the Greeks, earlier than by the Moors from Spain"; as a result, he says, in Germany cotton paper "was known in the 9th century by the name of Greek parchment."[62] That "Greek parchment" would be used to designate cotton paper highlights the recursivity of supersession narratives: the superseding object (paper in this case) makes sense of itself by reference to the presence or memory of the superseded object (parchment).[63] Indeed, Koops refers to manuscripts from tenth- to twelfth-century Spain that are "intermixed parchment and thick cotton-paper leaves." Scarcity is the reason for intermixing, he suggests, but he cannot determine if "cotton paper was scarcer than parchment, or that this mixture was necessary because sufficient parchment could not be obtained."[64]

The same sort of incomplete supersession happens with cotton and linen paper in Koops's account. According to *Substances*, the use of cotton paper in England "continued till the latter end of the 14th century," he says, before it was "gradually supplanted by linen-paper."[65] Here it is hard to miss the agricultural pun, intended or not, that comes into play with the word "supplanted." But beyond wordplay, there is a literal shift in the direction of replacement. To supersede something is to sit above it; to supplant something is to trip it from below. Cotton paper is, quite literally, tripped up from below by a new plant, flax. And yet, as Koops notes, the change is "gradual," not sudden. In Fabriano, one of the earliest and most well-known Italian papermaking cities, Koops claims that "cotton might have been some time before mixed with linen-rags, till the superiority of the latter was fully ascertained."[66] Koops devotes ample attention to teasing out when linen paper was invented, but it turns out the when is in service of the where: nationalistic disputes abound when it comes to assigning credit for the innovation of linen-based paper. Koops ultimately sides with the view that the earliest-known linen paper was made in Germany in 1308. "It is, therefore likely," writes Koops, "that Germany has the honour of its [linen paper's] invention."[67] By 1342, linen paper was in use in England, Koops claims, but it was not until the end of the seventeenth century that "the art of making Paper arrived to a great degree of perfection in England."[68] But within one hundred years, production speed (and demand) outstripped supply in England, and, as a result, English papermakers had to import rags from Europe, leading to anxieties about the dissemination of nationally beneficial information and even the production of viable banknotes. Koops raises the concern, in contradistinction to the anonymous eighteenth-century writer cited by Hunter, that due to resource shortage, there might not be enough rags to make paper money.[69]

Extraordinary Scarcity

As Koops had already lamented at the beginning of *Substances*, "The great demands for Paper in this country have rendered it necessary to be supplied from the continent. This supply is extremely precarious, and is likely to be more wanted, as the consumption of Paper increases, because the material, which is the basis of Paper, is not to be obtained in England in sufficient quantity. The evil consequence of not having a due supply of Rags has been the stoppage of a number of Paper-mills."[70] According to Koops, though "all Europe has of late years experienced an extraordinary scarcity" of rags, England struggled more than any other country because it relied on raw materials from abroad.[71] Parliament had already responded to rag scarcity by allowing rags, nets, old ropes, and waste paper to be imported duty-free. And yet, claims Koops, such measures were stopgaps, not solutions; in his words, they "cannot sufficiently obviate the lamentable scarcity."[72] Koops's appeal ultimately hinges on the argument that continued rag dependence is shortsighted, as are attempts to fix the problem economically (duty-free imports of raw materials), politically (sanctions on exports of certain materials to certain countries), and socially (appealing to citizens to be better stewards of rags).[73] The narrative Koops tells about the past is meant to be instructive: the history of substances used to convey ideas is one of supersession, and so is the future of substances used to convey ideas.

Koops was not the first to try to make paper from straw, and he does not claim to be. His claim, printed on the title page, is that his is the first *useful* paper made only from straw (with no rags added). By September 1800, the date given at the end of the preface to *Substances*, Koops had secured two patents related to his work on substitute materials for papermaking: one for extracting ink from printed and written paper, pulping the waste paper, and making new paper (No. 2392, April 28, 1800); and one "for a method of manufacturing paper from straw, hay, thistles, waste, and refuse of hemp and flax, and different kinds of wood and bark" (No. 2433, August 2, 1800).[74] His motive for printing *Substances* appears to have been twofold: to prove the success of his project and to appeal to the Crown to protect his trade secrets from foreign theft. By printing the book on straw paper, Koops demonstrates its viability. In numerous copies I have examined, Koops has signed the preface, perhaps to show that straw makes good writing paper as well (see Figure 6). In the latter pages of *Substances*, Koops claims that he has manufactured "several thousand reams of perfectly clean and white Paper, since the 1st of May, made from old waste, written, and printed Paper."[75] His next goal is to make "the most perfect Paper from straw and other vegetables." The paper "these lines are printed upon" is not yet perfected, he claims, but "this specimen leaves no doubt in my mind, that . . . I shall make straw-paper in as great perfection as that which is now re-manufactured from waste-paper."[76] And if the project fails? Koops makes the case that even if straw paper were only good for "pasteboards, packing-paper, and paper-hangings," it will benefit the nation.[77] According to Koops, making paper-hangings from English straw has the potential of flipping an upside-down economic model so that, instead of sending money to the Continent for rags, England would be able to manufacture paper-hangings more cheaply and could then sell paper-hangings at a lower rate to European purchasers. Straw paper,

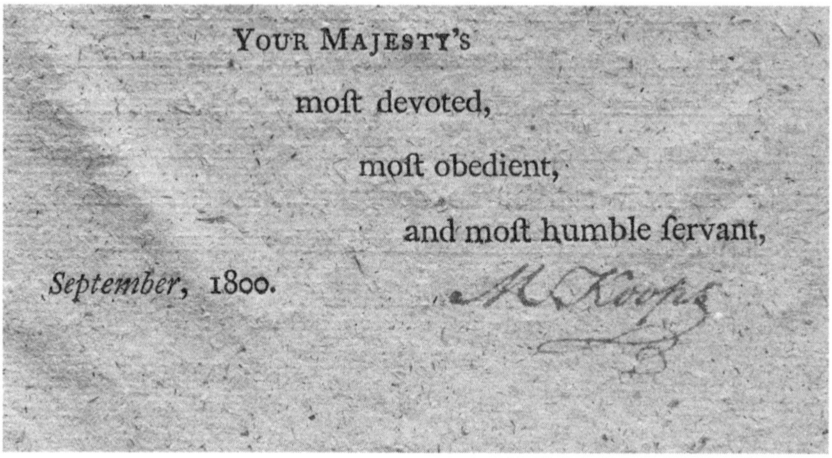

FIGURE 6. Koops's signature at the end of the preface to *Substances* (1800). Courtesy of the Department of Special Collections, Memorial Library, University of Wisconsin–Madison.

Koops argues, is a win-win proposition that can drastically reduce dependence on foreign supply by substituting locally sourced (or, at least, nationally sourced) raw materials.

At the end of *Substances*, Koops adds a brief, seven-page (two sheets with seven of eight possible sides printed) appendix, which is posed as if it is an afterthought, but stands out to a reader because the paper on which it is printed is clearly not the same as the paper used for the rest of the book. The sheets are more pale brown than yellow, and they are less coarse. Printed on this paper are these words: "The following lines are printed upon Paper made from Wood alone, the produce of this country, without any intermixture of rags, waste paper, bark, straw, or any other vegetable substance."[78] Koops had figured out the future of papermaking. He had successfully produced and printed on paper made from wood alone, with no additives. And the paper remains intact in all of the editions I have examined.[79] Even in 1801, when Koops expanded *Substances* and printed the main part of some editions on "Paper Re-made from Old Printed and Written Paper" instead of on straw paper, the wood-pulp paper appendix is still present. However, only in the 1800 edition do we find, at the end of the volume, a small ornament that might well be an emblem of *Substances* as a whole: the image of a gravestone, engraved with the word "FINIS" in front of a chopped-up tree and a setting sun.[80] The future of paper, as Koops figured out, was to be found in the harvesting of trees.

Looking backward from 1800, Koops sees how bookmaking brought the world to the cusp of the Industrial Revolution; looking forward, he sees the

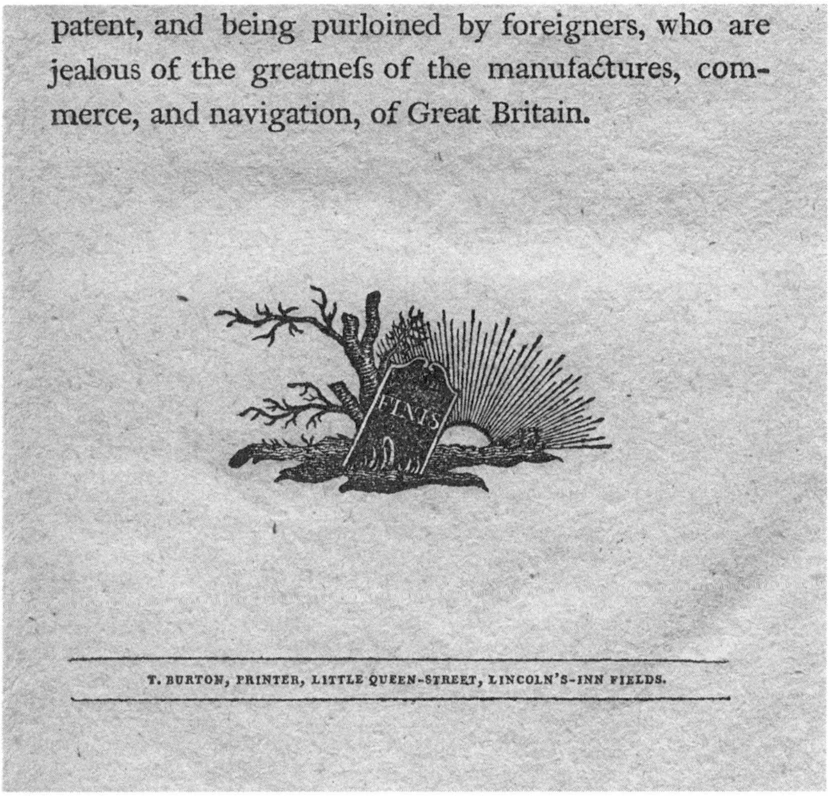
patent, and being purloined by foreigners, who are jealous of the greatneſs of the manufactures, commerce, and navigation, of Great Britain.

FINIS

T. BURTON, PRINTER, LITTLE QUEEN-STREET, LINCOLN'S-INN FIELDS.

FIGURE 7. Ornament from the last page of *Substances* (1800), printed on wood pulp paper. Courtesy of the Department of Special Collections, Memorial Library, University of Wisconsin–Madison.

possibility of establishing the manufacture of paper in England using a substance, straw, that is available in a sustainable, abundant supply. The book itself is addressed to King George III and framed as an appeal about licensing and patents more than as a persuasive work of book history scholarship. And yet it is both. But whatever his talent for understanding the history of papermaking, Koops's own project failed spectacularly. By 1800, he had secured three patents for papermaking—all three for experimental approaches to replacing rags with more abundant raw materials—and he was working at Neckinger paper mill in Bermondsey. By 1801, Koops had established the Straw Paper Manufactory at Millbank in Westminster, the largest paper mill in England. By October 1804, Koops was bankrupt, and the mill's equipment was on the auction block, "ending the possibility of England challenging the European paper industry by using more easily available materials for making paper."[81]

Koops's narrative unsettles the sort of hindsight biases that do not often acknowledge that books are equal parts human ideas and nonhuman materials. The brilliance of Koops's volume is that it simultaneously demands several kinds of reading, for its pages—the words on them and the paper itself—record a natural history of the book and, especially, of papermaking. When Koops writes about pages made from raw materials, he writes not about substrates, but about substances; not about inevitability, but about potential; not about books or media, but about memorials; not about ecological abundance, but about ecological scarcity. The irony is that Koops ushers papermaking into the industrial age of mass production.[82] The coarse yellow pages of *Substances*, the pale brown pages of its wood-pulp appendix, and the story that Koops tells on those pages mark the end of an epoch of hand papermaking. Within a decade of *Substances* and of the failure of Koops's paper mill, papermaking would be revolutionized (and industrialized) with the help of bleach and the perfection of papermaking machines. Raw materials would remain problematic and scarce, but by the late nineteenth century, the papermaking industry shifted westward, across the Atlantic Ocean and across the North American continent, as trees became the cheap and reliable plants that could be transformed into paper.[83]

The untapped resources of the American West inspired belief in the obviousness of what progress was and in a providential mandate to subdue and have dominion. Reproductions of *American Progress* were not included in nineteenth-century guidebooks to temper optimism. The story of paper is a similarly complicated narrative that, in many histories of paper, easily veers into encomium. One particularly adulatory history of paper imagines an uncomplicated, machine-driven paper utopia: "In less than half a century, the machines have entirely superseded the diminutive hand-mills which sparsely dotted the country, and gigantic establishments have risen up in their places. Paper-mill villages, and banking institutions even, have grown out of this flourishing branch of industrial art, and we behold with satisfaction and amazement, what has been brought about by the aid of a commodity so insignificant in the eyes of the world as linen and cotton rags."[84] This is the *Chronology of the Origin and Progress of Paper and Paper-Making*, printed four years after Gast's *American Progress*, and a fitting companion to it, for the story of early handmade paper is not pre-industrial, but proto-industrial. It is a story of questionable progress toward cheap raw materials and cheap labor—"cheap at any cost," to cite Wendell Berry.[85]

Spinning Wheels

Compared to Taylor's seventeenth-century poem of praise for hemp and flax and to Thoreau's paean to the commodities moving across nineteenth-century train tracks and even to Darnton's far more recent Communications Circuit, Koops offers a striking perspective on the scarcity and abundance of substances used to record ideas. Reading Taylor and Thoreau, one could be forgiven for assuming that raw materials were always there—somewhere. The second half of Taylor's *Praise of Hemp-seed* celebrates paper made from hemp and flax rags as a "rich commodity" worthy of excessive praise because, as Taylor puns, "all the world yeelds matter to my Muse." That is, plant matter in the form of cellulose from around the globe yields paper, which in turn yields poetic matter about the global importance of paper. Taylor goes on (and on):

> No Empire, Kingdome, Region, Prouince, Nation,
> No Principality, Shire, nor Corporation:
> No country, county, city, hamlet, towne,
> But must vse paper, either white or browne.
> No Metropolitane, or gratious Primate
> No village, pallace, cottage, function, climate,
> No age sex, or degree the earth doth beare,
> But they must vse this seed to write, or weare.[86]

Reading Taylor, paper users must have known the story of paper made from plants was not so simple, direct, or domestic as his commendatory lines suggest. Perhaps Renaissance writers and readers knew about the fraught supplies of foreign paper the way we know as a culture, but do not operate as though we believe, that we will not always be able to rely on cheap, foreign fossil fuels. In an imaginary dialogue about the state of the country, printed in London in 1581, one of the interlocutors explains that "there was paper made a while within the realme," but that the papermaker could not make his paper "as good cheape as it came from beyond the sea."[87] The papermaker referred to was probably John Tate, the first papermaker in England around the end of the fifteenth century, but the complaint is the same made by Matthias Koops, the first straw and wood pulp papermaker in England at the beginning of the nineteenth century.

When Taylor considers the origins of paper, how a plant like hemp (or flax) cycles through society and then "to paper doth conuert," he is drawn to think about how human stories and plant stories come to be interwoven:

> For when I thinke but how is paper made
> Into Philosophy I straight waies wade:
> How here, and there, and euery where lies scatter'd,
> Old ruind rotten rags, and ropes, all tatter'd.[88]

Taylor goes on to imagine several specific origin stories in which paper becomes a rhetorical under-text that inflects and in some cases subverts the over-text. A Brownist's "zealous ruffe," for instance, might "Be turnd to paper, and a Play writ in't."[89] But the most famous lines that Taylor records in this poem imprinted into the fibers of plant-based paper are these:

In paper, many a Poet now suruiues
Or else their lines had perish'd with their liues.
Old *Chaucer*, *Gower*, and Sir *Thomas More,*
Sir *Philip Sidney,* who the Lawrell wore,
Spencer, and *Shakespeare* did in Art excell,
. . . .
Forgetfulnesse their workes would ouerrun,
But that in paper they immortally
Do liue in spight of death, and cannot die.[90]

Taylor's lines constitute the first posthumous mention of Shakespeare's name in print.[91] As Adam Hooks notes, Taylor gives precedence to "the material and economic context": Shakespeare defies death "because individual copies of his plays and poems were printed, bought, and sold."[92] Shakespeare, like the other poets listed, survives "in paper"—a claim that subtly plays with Shakespeare's own observation in *Sonnet* 15 that "men as plants increase."[93] As I discuss in the next chapter, the connection between poet and plant is not chiefly metaphorical but material, and it is recognizable in the handmade pages of early books in the botanical form of visible flecks of plant matter.

Thoreau's paper, by comparison, appears dull. Though still composed of rags, it is "wove" paper made by machine, and the pages of the journals he kept while at Walden Pond tend toward homogeneity.[94] Mark Bland writes that "modern paper"—with which Thoreau's paper shares many visual characteristics even though the nature of his pages shares more ecological characteristics with Taylor's and with Shakespeare's paper—"is mass produced and production methods seek to minimise differences . . . in effect, the paper either effaces or standardises the history within it."[95] At the Morgan Library, where I looked into the pages of some of Thoreau's journals along with curator and paper historian John Bidwell, I became especially aware of how much can be effaced in the homogeneity of modern paper. I was struggling to read Thoreau's terrifically scribbled handwriting and defaulted to the most basic of paleographical skills: striving to positively identify unique letterforms as a basis for decoding full words. On the page shown in Figure 8, one letterform immediately stood out toward the middle, left margin, a capital letter that had to be a *Z*. And suddenly,

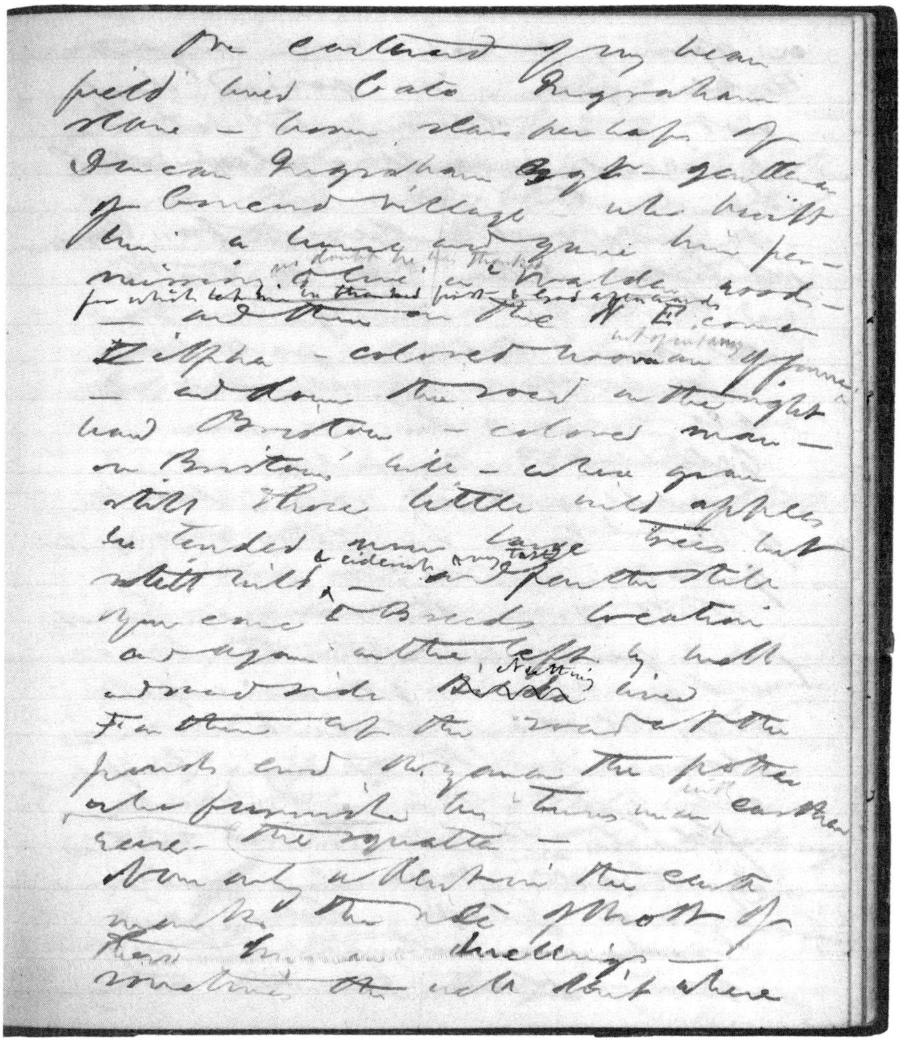

FIGURE 8. Page from Henry David Thoreau's journal while at Walden Pond. MA 1302.7. Thoreau, Henry David, 1817–1862. Journal: autograph manuscript: Walden [Concord], [1845 July–Winter 1845/1846], p. [87] "Over eastward of my bean field. . . ." Reproduced by permission of the Morgan Library and Museum.

out of the mire of Thoreau's scrawling on recycled plant-fiber rags, Zilpah White emerged. Zilpah White, the former slave whose Loyalist master abandoned her to fate when he fled the country in 1775.[96]

White had gone to Walden Woods not to suck the marrow out of life, but to survive independently, and she did so with a flax wheel. In *Black Walden: Slavery and Its Aftermath in Concord*, Elise Lemire explains that White's former masters had a "spinning garret" for "spinning flax into the linen fibers that were then woven into fine table linens and other markers of wealth and gentility."[97] Thoreau, telling White's story decades after her death, writes that "she spun linen for the townsfolk, making the Walden Woods ring with her shrill singing."[98] Zilpah White's story is inextricably connected with the story I am telling about paper, and the role she played in the history of plants as rags as paper is largely hidden from view. "When slavery ended," writes Lemire, "at least some former slave women were given one of their master's spinning wheels as a means of providing for themselves."[99] It was a short-term solution, however, for the work these women could do on small spinning wheels would soon be overwhelmed by industrial factories that could make linen cloth more quickly and more reliably and sell it more cheaply.[100]

Paper has a serendipitous and romantic and compelling and complex backstory, one that can begin to seem so obvious in hindsight, one that marches across the pages of history like Columbia across the bountiful landscape of the American West. Rag shortage becomes a useful foil in many histories of paper, for it creates dramatic tension while simultaneously introducing human-interest stories of ingenuity, pluck, and thrift. Rag shortage is certainly one part of the story, but an overemphasis on rags has made it easy to forget that rags, like books or clean drinking water or oil or cucumbers, come from somewhere and something. It is true that England lagged behind the rest of Europe when it came to paper production and "relied on imported paper for almost all of its writing and printing needs before the end of the seventeenth century."[101] It is true that when Richard Tottell, printer of law books and of the first anthology of English poetry and charter member of the Stationer's Company, petitioned for exclusive rights to papermaking in England in 1585, he "accused the French of cornering the supply of rags."[102] But it is also true that, as D. C. Coleman observed over sixty years ago, English rag shortage could be attributed to three factors: "the widespread use of wool . . . the absence of a native linen industry . . . and to the fact that such linen rags as were available tended to be exported, particularly to France."[103] Stories about paper have long focused on the human elements in the unbalanced rag-shortage equation Coleman describes: wool use, linen industry, rag exports. However, when we read between the lines—and beyond the linen—we realize that "rag shortage" is a euphemism for raw material shortage. Hidden below the canopy of the familiar, anthropocentric narrative in which "rags make paper" are dense understories of raw materials, of localized biodiversity, of bruised plant stalks and boiled animal hooves and of aching, dehumanized bodies. If we look closely, we begin to find stories in the fibers of historical paper that prompt us to recalculate the costs of supposed cheapness.

CHAPTER 2

The Word Made Flax: Cheap Bibles, Textual Corruption, and the Poetics of Paper

> A handmade sheet of paper provides documentation or physical evidence of a dialogue between human beings and nature.
>
> —Timothy Barrett, "Aesthetics and the Future of the Craft"

In a controversial history of *The Norton Anthology of English Literature*, Sean Shesgreen claims that one of the three reasons for the anthology's remarkable success in the 1960s and thereafter was its so-called innovational use of "Bible paper."[1] According to Shesgreen, "Bible paper had previously been shunned by anthologists: transparent and flimsy, it tears easily and bleeds profusely."[2] But the use of "Bible paper" allowed *The Norton Anthology* to offer 60 percent more content than its main competitor in a text that weighed 25 percent less, and more pages with less bulk allowed the printers to use a smaller size (octavo-size rather than quarto-size).[3] The new, flimsy-paper anthology was an unprecedented success, but that success might be qualified by considering the degree to which substandard paper ultimately affected student perceptions of the literature printed on that paper. What message did the new media communicate? Is it possible that the "flimsy" paper emphasized functionality over form, coding its printed contents as means to an end (an acceptable course grade, for example) rather than as avenues of aesthetic exploration? In short, what is the rhetorical effect of cheap paper, especially as a medium for supposedly cherished literature?

Historically and conceptually, the anthology's bookmaking innovation was hardly innovative. The flimsy paper is called "Bible paper" for a good reason: Bibles have a long history of being printed on cheap paper, as *The Norton Anthology*'s editors must have known. The trick of cutting paper costs to make a Bible more portable was used even before paper and printing: "Paris Bibles" from the thirteenth century were copied by secular scribes onto "tissue-thin parchment"[4] to create a portable Bible "intended to meet the needs both of the student in the university classroom and the Mendicant preacher on a mission."[5] Printing-press technology itself developed only after papermaking technology reached Europe, bringing with it a cheaper alternative to parchment.[6]

In *Print and Protestantism in Early Modern England*, Ian Green demonstrates that "canny publishers used nearly every trick in the book to expand markets and maximize profits" on Bible sales, including the use of "cheaper paper."[7] As with Shesgreen's discussion of Bible paper, scholars have focused on the increased portability, distribution, and ownership of cheaper Bibles. What tends to be overlooked are the aesthetic effects of the surfaces on which words appear. This oversight is particularly remarkable for three reasons. First, since current textual criticism has been deeply influenced by D. F. McKenzie's work on the "sociology of texts" and the view that "forms effect meaning," we might expect to read more about paper, one of the book's most basic forms.[8] Second, paper is one of the formal features most legible to historical readers, especially the seventeenth-century readers I discuss here, who may not have understood the nuances of the printing press, but who participated in and understood the nuances of the rags-to-paper economy. Third, as I will show, these same historical readers actively commented on the aesthetic effects of paper quality.

Conversations (and controversies) about cheaply produced Bibles range through the dates and geographies covered in *The Norton Anthology*, but in this chapter I consider the ways that the Protestant Reformation made the Bible—and, by extension, other books—more vulgar, to use a term deeply associated with Bible translation and transmission. The Reformation's doctrinal emphasis on personal reading and interpretation dramatically increased book ownership and literacy rates in Renaissance England as the Bible was made to be both physically and intellectually grasped by readers-in-training.[9] John Dryden famously decries the material effects of the Reformation: "The Book thus put in every vulgar hand, / Which each presum'd he best cou'd understand, . . . / The tender Page with horney Fists was gaul'd."[10] As Dryden's critique suggests, physical graspability had interpretive consequences. Margreta de Grazia writes, "If words are to serve as transparent representations of things, their own thing-like or sensible properties must be overlooked."[11]

For de Grazia, "the material properties of words" are "their duration as sound when spoken and their extension as marks when written."[12] To the thing-like properties of sound and symbol, I add texture. In this chapter, I argue that the texture of words, and the texture, especially, of the media on which "God's words" were printed, were not effectively overlooked by Renaissance readers, who had the capacity to recognize the things from which their texts were constructed. I move from my focus on the materiality of paper as a transformed plant in Chapter 1 to focus here on the poetics of paper in Renaissance English texts and more broadly on what I would call a poetics of corruptibility. Bibles themselves were byproducts of organic life cycles, of germinated seeds and rotting plant stalks. Recognizing the ecology of the book as a Renaissance reader might have, we might grasp—both literally, as page texture, and cognitively, as aesthetic insight—a richer, more poetically intriguing interplay between recorded words and the decaying substrates on which they appear.

I begin with a discussion of paper quality and printing costs in Renaissance England that focuses on the books that most influenced literacy and reading practices in Renaissance England: vernacular Bibles. Though there are many voices and opinions in the debate over words as things in Reformation-era England, this chapter is guided by one particular conversant, Henry Vaughan (1621–95), who, while reading his "cheap" Bible, is conscious of the intersecting lives of bookish words and natural matter in the past, present, and future.[13] Vaughan offers a palimpsestic reading strategy that anticipates our own critical turn toward "polychronic" readings as articulated by Bruno Latour and Michel Serres, and as convincingly applied to Renaissance literary criticism in Jonathan Gil Harris's *Untimely Matter*.[14] Vaughan's poem "The Book," first printed in the second edition of *Silex Scintillans: Sacred Poems and Private Ejaculations* in 1655, relies upon a readerly understanding of the social cycles of flax plants, among other things, as he itemizes the flora and fauna used to make his Bible:

The Book.

Eternal God! maker of all
That have liv'd here, since the mans fall;
The Rock of ages! in whose shade
They live unseen, when here they fade.

Thou knew'st this *papyr*, when it was
Meer *seed*, and after that but *grass*;
Before 'twas *drest* or *spun*, and when
Made *linen*, who did *wear* it then:
What were their lifes, their thoughts & deeds
Whither good *corn*, or fruitless *weeds*.

Thou knew'st this *Tree*, when a green *shade*
Cover'd it, since a *Cover* made,
And where it flourish'd, grew and spread,
As if it never should be dead.

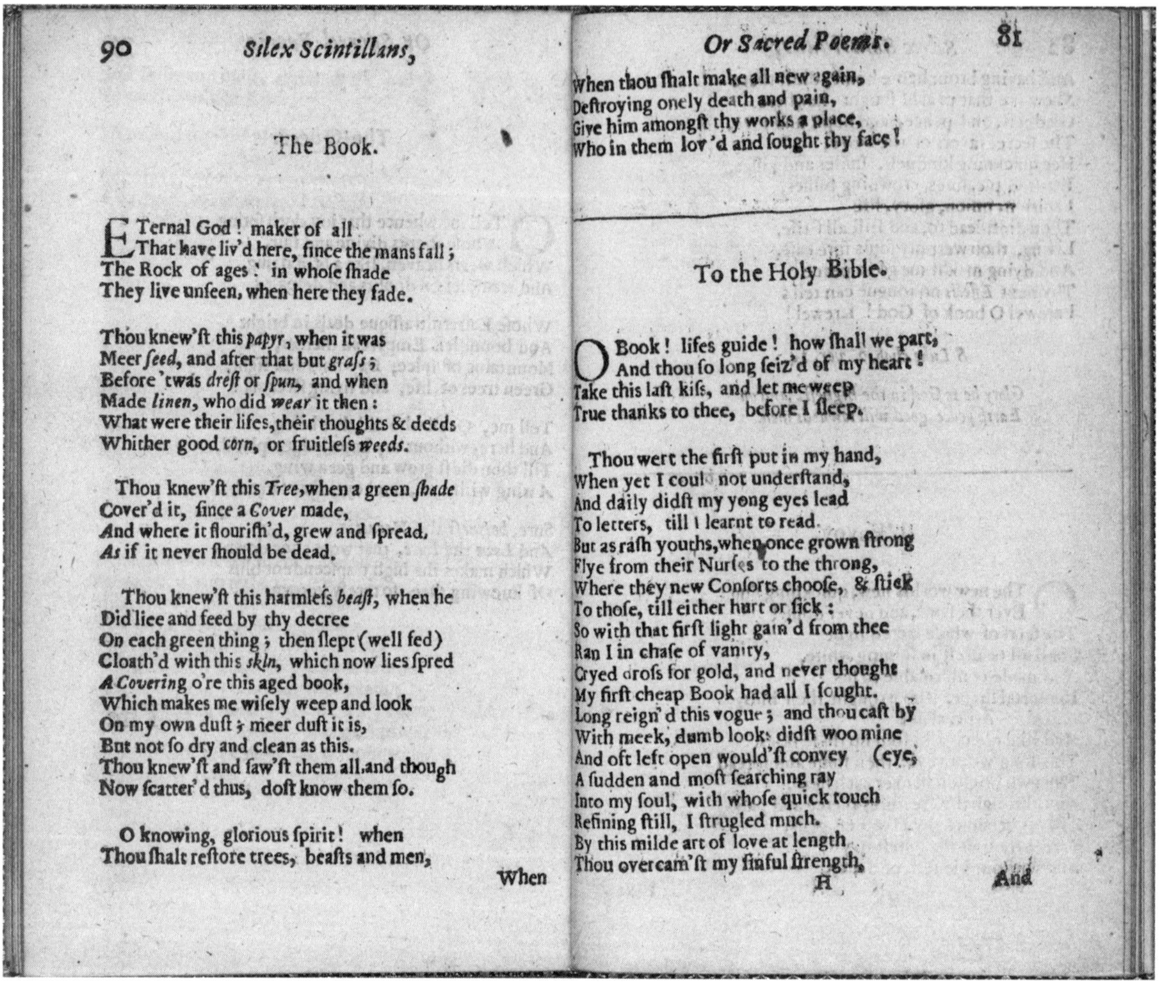

90 *Silex Scintillans,*

The Book.

ETernal God! maker of all
That have liv'd here, since the mans fall;
The Rock of ages! in whose shade
They live unseen, when here they fade.

Thou knew'st this *papyr*, when it was
Meer *seed*, and after that but *grass*;
Before 'twas *drest* or *spun*, and when
Made *linen*, who did *wear* it then:
What were their lifes, their thoughts & deeds
Whither good *corn*, or fruitless *weeds*.

Thou knew'st this *Tree*, when a green *shade*
Cover'd it, since a *Cover* made,
And where it flourish'd, grew and spread,
As if it never should be dead.

Thou knew'st this harmless *beast*, when he
Did liee and feed by thy decree
On each green thing; then slept (well fed)
Cloath'd with this *skln*, which now lies spred
A Covering o're this aged book,
Which makes me wisely weep and look
On my own dust; meer dust it is,
But not so dry and clean as this.
Thou knew'st and saw'st them all, and though
Now scatter'd thus, dost know them so.

O knowing, glorious spirit! when
Thou shalt restore trees, beasts and men,
When

Or Sacred Poems. 81

When thou shalt make all new again,
Destroying onely death and pain,
Give him amongst thy works a place,
Who in them lov'd and sought thy face!

To the Holy Bible.

O Book! lifes guide! how shall we part,
And thou so long seiz'd of my heart!
Take this last kiss, and let me weep
True thanks to thee, before I sleep.

Thou wert the first put in my hand,
When yet I could not understand,
And daily didst my yong eyes lead
To letters, till I learnt to read.
But as rash youths, when once grown strong
Flye from their Nurses to the throng,
Where they new Consorts choose, & stick
To those, till either hurt or sick;
So with that first light gain'd from thee
Ran I in chase of vanity,
Cryed dross for gold, and never thought
My first cheap Book had all I sought.
Long reign'd this vogue; and thou cast by
With meek, dumb looks didst woo mine
And oft left open would'st convey (eye,
A sudden and most searching ray
Into my soul, with whose quick touch
Refining still, I strugled much.
By this milde art of love at length
Thou overcam'st my sinful strength,
H And

FIGURE 9. "The Book," in Henry Vaughan's *Silex Scintillans* (1655), GEN 14456.49.2.10*, G8v–H1r, Houghton Library, Harvard University.

Thou knew'st this harmless *beast*, when he
Did liee and feed by thy decree
On each green thing; then slept (well fed)
Cloath'd with this *skln* [*sic*], which now lies spred
A *Covering* o're this aged book,
Which makes me wisely weep and look
On my own dust; meer dust it is,
But not so dry and clean as this.
Thou knew'st and saw'st them all and though
Now scatter'd thus, dost know them so.

O knowing, glorious spirit! when
Thou shalt restore trees, beasts and men,
When thou shalt make all new again,
Destroying onely death and pain,
Give him amongst thy works a place,
Who in them lov'd and sought thy face![15]

Vaughan's poem, discussed in greater detail at the conclusion of this chapter, converses with sixteenth- and seventeenth-century debates about cheap media and the production of a vernacular Bible in England, and it addresses modern critical concerns about material culture, the "thingness" of words, and the sociology of texts. What is perhaps surprising for a twenty-first-century reader is Vaughan's detailed awareness of the natural resources used to make his Bible, particularly his expectation that readers will understand his references to animal and plant origins. What is more surprising, though, is the way that Vaughan *reads* these natural resources and their potential for corruption, especially in light of the early history of vernacular Bible production in England. "The Book," arguably "one of [Vaughan's] most impressive" poems, is pleasing as a poetic conceit, a conversation between a poet and the "maker of all" about the organic nature of Renaissance books.[16] As I will show, it is also, on a deeper level, a contribution to an ongoing cultural conversation about the poetic form and function of the natural things on which "the Word" was printed in Reformation England.

Re-Forming the Bible in Renaissance England

By the turn of the seventeenth century in England, as I have noted, vernacular Bibles were becoming cheap. They were made to be affordable, and, as a result, book ownership and literacy rates in England spiked. William H. Sherman claims that the Geneva Bible alone, "which went through more than 140 editions between the 1560s and the 1640s," was probably "the most widely distributed book in the English Renaissance, and the one that played the most crucial role in changing the patterns of lay book ownership in the age of print."[17] In Renaissance England, according to Sherman, "literacy did not mean just reading; it meant *reading the Bible*."[18] The translation of God's Word to mass media posed material challenges, and some were aghast at the relative cheapness of Bibles. William Prynne writes in *Histrio-mastix* that some playbooks "are growne from *Quarto* into *Folio*; which yet beare so good a price and sale, that I cannot but with griefe relate it, they are now new-printed in farre better paper than most

Octavo or Quarto *Bibles.*"[19] A printed marginal note adds: "Shackspeers Plaies are printed in the best Crowne paper, far better than most Bibles." Prynne's complaint about the paper quality of Bibles in relation to the quality of Shakespeare's First and Second Folios suggests that the relative cheapness of Bible materials was inherently aesthetic. John Vicars, in a 1645 pamphlet, appeals for a "Reformation" of the "sacred Book of God" and rails against the printing of Bibles filled with errors "wherby the *sense* is in very many places, *feully* corrupted and *falsly* mistaken" ("foully," misspelled as "feully," ironically reinforces Vicars's point).[20] Vicars further criticizes the "*printing of our Bibles* in very course and extreme thin and bad *sinking-paper.*"[21] In Vicars's point of view, Bible printing is a neglectful business that seeks after "filthy lucre" by using bad materials.[22] These materials, however, are not value-neutral, for they tend to affect Bible reading; errata may lead a reader astray, and "*sinking-paper,*" paper that is not well coated with gelatinous animal sizing (a topic I discuss in detail in Chapter 4), allows water-based handwriting ink to blot into its fibers, frustrating a reader's attempts to annotate the margins of his or her Bible.[23] Edward Leigh similarly condemns the use of cheap paper in Bible printing. His list of those who "unreverently handle the Scriptures" includes the usual suspects, such as atheists, papists, antinomians, and witches.[24] Less predictably, printers "who print the Bible in bad Paper" make the list ahead of heathens and Jews.[25]

According to Green, "The rapid expansion and diversification of bible production in England . . . was due primarily to a combination of God and Mammon."[26] Green details English "publishers' success in devising and disseminating cheaper, simpler bibles" to meet market demands.[27] This "diversification of formats," as Green demonstrates, was achieved not only by making Bibles increasingly smaller, but also by manipulating paper quality.[28] Green shows evidence of folios being sold in three qualities of paper (superior, fine, and ordinary) and of quartos and octavos sold in two qualities of paper. Of course, cheaper production costs did not always mean that discounts were passed on to customers. In *Scintilla, or, a light broken into darke Warehouses,* Michael Sparke denounces the "Monopolists" who were manipulating Bible prices.[29]

Sparke's six-page pamphlet is filled with references to "quires"—used twenty-three times—and to relative paper quality: "thinne paper," "good paper," "Large paper," "better paper," "best paper," "excellent paper."[30] In one instance, Sparke complains that "*Church* Bibles of a thinner sort" were "cheaper" and "were excellent for poore Parishes" until the monopolists exploited the market.[31] Sparke's diatribe suggests the value of attending to what we might call the aesthetics of cheapness in relation to texts and especially paper. As Green notes, though there is a lack of reliable pricing data for sixteenth- and seventeenth-century Bibles, one can confidently discuss the *relative* cheapness of particular works and the rhetorical effects of the varied paper qualities discussed (and denounced) by Prynne, Vicars, and Sparke.[32]

It is worth noting, too, that from leaf to leaf within printed texts, paper usage was not always consistent. According to David McKitterick, "Although printers usually endeavoured to ensure reasonable continuity of quality and colour throughout a volume, there were exceptions," including a 1648 text printed on "stocks varying in colour between shades of brown and white," a 1642 folio Greek New Testament printed on different paper sizes, and a 1629 Bible composed of "no less than seven different stocks, divided between discrete issues."[33] One must also consider the frequency with which discrete volumes were bound together, putting the variant paper qualities of Bibles and Bible aids (prayer books, metrical Psalms, concordances, etc.) in juxtaposition. The expense of binding "meant that works rarely stood alone in self-enclosed units but were mixed with other works to save money."[34] It is easy to overlook the page and thus miss the significance of mixed paper stocks. Or, noticing varying qualities of paper from title to title within a *Sammelbände* or even within individual titles, it is easy to settle for an explanation grounded in the economics of the book trade. But the book trade exists in a world of things, and the mixing of paper qualities is not just the mixing of colors and grades of paper but also the mixing of species. Varied paper qualities and mixed paper stocks draw attention to the ecologies of Renaissance papermaking; at the same time, they draw attention to the aesthetic perceptions of Renaissance readers who commented upon the substances used to convey ideas.

For example, Leigh, who criticizes Bible printers for using bad paper stock, also repeats a material criticism of the printed Bible: "The Papists stile the Scripture . . . the black Gospell, inky Divinity."[35] When a printed Bible is accused of being a "black Gospell" and an "inky Divinity," part of the criticism is that the communication of godly ideas is muddled in the process of textual production. For Catholics, the Bible is supplemented by unwritten tradition, the direct, unmediated communication of God with the clergy. When the Bible is attributed sole authority, it becomes, in the view of Catholic critics, an idolatrous divinity made of ink. Or, materially, an idol made of flax, for the "basic constituent" of English printing ink in the period was linseed oil.[36] The Word made flax, in the Catholic view, is the Word corrupted beyond recognition. "Inky Divinity" is vibrant matter of the worst kind, a blotty collation of matter that damns the souls of humans.

Debates about the vernacular Bible in England, which centered on the issue of *sola scriptura*, were thus complicated by the practical means by which the Bible was made available to lay readers. In short, the Protestant Reformation and *sola scriptura* gave rise to Bibles made cheaply of cheap materials. Bibles were more tangible, more accessible, and more handy. Their margins were often markable, and readers were encouraged to write in them. Reformation Bibles were secular; what had previously been hand-copied by religious scribes was being mass-produced by merchants—with sometimes ghastly results, such as "The Wicked Bible," a 1631 Authorized Version whose seventh commandment read, "Thou shalt commit adultery."[37] Implicit in period debates about the Bible, the text that transformed reading practices in Renaissance England, is a debate about its material status as a set of metaphysical ideas bound up in the physical world.

Words on Flax

Thomas More and William Tyndale, the early voices in debates over the introduction of vernacular Bibles in England during the reign of Henry VIII, fundamentally disagreed about the role of the "unwritten," that is, of the mediating role of church tradition.[38] Tyndale believed in the authority of the Bible alone, *sola scriptura*, rejecting the authority of the church's oral tradition. For More, the biblical text was not the only source of revelation, and he rejected the grounds on which Tyndale wished to argue. As Germain P. Marc'hadour and Thomas M. C. Lawler point out, "More's response [to Tyndale] . . . is not to quibble over the text of scripture but to interpret the sense of the passage in terms of the tradition of the church."[39] English translation (or mistranslation) was certainly a point of contention between More and Tyndale, but it is one that has received more attention than the more basic disagreement about textual authority.[40] Essentially—or, rather, substantially—More and Tyndale argued over media and corruption. Tyndale distrusted the clergy as unreliable mediators between God and humans. Power corrupts. More, on the other hand, distrusted textual media as unreliable, citing as evidence the facts that (1) some scripture has already been lost, (2) we cannot know how much, and (3) parts of what we have are "corrupted."[41] Texts tend toward corruption, and that corruption is due not only to human error, but also to material conditions. Paper is easily ripped, burned, and soaked. Bookworms are no respecters of crucial words, and knots of organic matter in the page can interrupt typography.

Modern scholars in archival libraries can still see plants in paper. That is, we can still easily see what Vaughan saw when he looked at his Bible and saw visible evidence of flax cultivation. The page space around and between words, often referred to as "white space," is anything but white. Though often assumed to be discolored by age, many of the brownish pages we encounter in archival libraries have actually retained coloring from their production. For instance, the rivers that provided water for paper mills were not always pristine, especially in the spring (when they ran muddy) or when the riverbanks were populated upstream. The "stuff" vat, the technical name for the pot of macerated fibers used to make sheets of paper, was about 99 percent water; so it is no surprise that silty, muddy, or polluted water would render sheets of paper darker.[42]

Looking a bit more closely, one can immediately see that the general impression of brownness is often enhanced by a network of flecks and fibers embedded in the page. They are ubiquitous in early texts to the point that they have become invisible to us. There are notable exceptions, however. Though most hairs in the pages of books are miniscule, and one has to look closely to see them, I find it hard to ignore long strands of hair that wind through the margins or printed areas of Renaissance books, even when the text on the page is not poeticizing the locks of a beloved. A feather embedded in a page of George Gascoigne's *A Hundreth Sundrie Flowres* is even harder to overlook; its placement in the margin points up a self-blazoning by the mourning lover in Dan Bartholomew.[43] Are we licensed to read this into the lines? Or to comment on the appropriateness of a feather appearing here, in a volume so preoccupied with textual production and misproduction, with material integration and disintegration? We can agree that unique instances of human marginalia in Gabriel Harvey's copy of Gascoigne are noteworthy and are representative of, or at least pertinent to, broader patterns of Renaissance book interactions; might we not argue the same for unique instances of avian marginalia?

Hairs are more common than feathers, but vegetable matter of two varieties is most commonly embedded in paper: (1) bits and pieces of vegetable fibers that made their way into clothing (during flax processing) and finally into paper, and (2) flecks of organic matter, presumably from riparian flora upstream, that were too small to filter out of the papermaker's vat. (As with the silty brownness described above, seasons and weather patterns affect the flecking of paper.) When one searches for readers' marks in books, flecks of organic matter can seem purposeful at first glance; in books where readers have used small, marginal tick marks as the primary method of highlighting passages, it is sometimes hard to tell (particularly in microfilm or digital reproductions) if a mark is made by pen *on* the paper or if the mark is organic matter *in* the paper. Occasionally a "knot" of organic matter embedded in the page will cause errors in the printed text, interrupting the type. Sometimes a large piece of flax appears in paper, a husky piece of the stalk of the flax plant known as a "shive."[44] An opening from a copy of the second edition of Hoby's translation of *The Book of the Courtier* is particularly illustrative of the ecologies of textual production and consumption, and shows evidence of hair interrupting the impression of the typeface of a printed marginal note on the verso, a shive extending off the trimmed edge of the page from the margin of the recto, and a brown, iron gall ink manuscript note below the shive.[45]

The largest, most noticeable shive I have found, in a copy of the first printing of More's collected English writings (1557), bears the ink of a printed marginal note. Shives and husks of flax are supposed to be eliminated in the process of converting plant to clothing. One might make the argument that while some bibliographers prefer to focus on textual accidentals and substantives in the printed area, actual readers may have been much more knowledgeable about this type of "ecological accidental." There is no human intention to be uncovered here, but the unexpected visibility of the world of things on which words were printed in Renaissance England yields an ecologically and poetically playful juxtaposition: the husk of flax interrupts a printed marginal reference to John 14, calling attention to physical substance beneath the citation of a passage in which Jesus explains the metaphysics of his incarnation. More striking, even, than bits of flax are swatches of cloth embedded in paper, like the approximately 12 x 12 millimeter swatch in a copy of Samuel Rutherford's *Lex, Rex: The Law and the Prince*.[46] Appearing, appropriately, in "The *Table* of the *Contents* of the *Book*," the texture of the fabric, with threads unravelling into the text block, emphasizes the rag content of the book's pages. We will never know if a remarkable flax shive or an unpulped rag in the pages of Vaughan's Bible was the inspiration for his poem "The Book." What we do know is that Vaughan, like his contemporaries, comprehended the natural origins of paper and understood that flax was literally inhabited—broken in as clothing—before it was used in papermaking.[47]

Discussing bookworms in *Micrographia, or, Some Physiological Descriptions of Minute Bodies Made by Magnifying Glasses with Observations and Inquiries Thereupon* (1665), Robert Hooke also points to the presence of husks of flax visible in paper: "This Animal [the bookworm] probably feeds upon the Paper and covers of Books, and perforates in them several small round holes, finding, perhaps, a convenient nourishment in those husks of Hemp and Flax, which have pass'd through so many scourings, washings, dressings and dryings, as the parts of old Paper must necessarily have suffer'd; the digestive faculty, it seems, of these little creatures being able yet further to work upon those stubborn parts, and reduce them into another form."[48] Wormholes, in Hooke's summary, are further reminders of the organic nature of the page and of the other "form[s]" that flax has and can become. The bookworm quite literally cuts out the privy use of book pages as toilet paper, leaving flax behind in "another form" and tempting one to comment on that which some books are full of. The intentional fallacy is hard to avoid. Goosy, hard-to-manage pens and page-scattering winds and blotchy ink used for Protestant Bibles are all imagined by

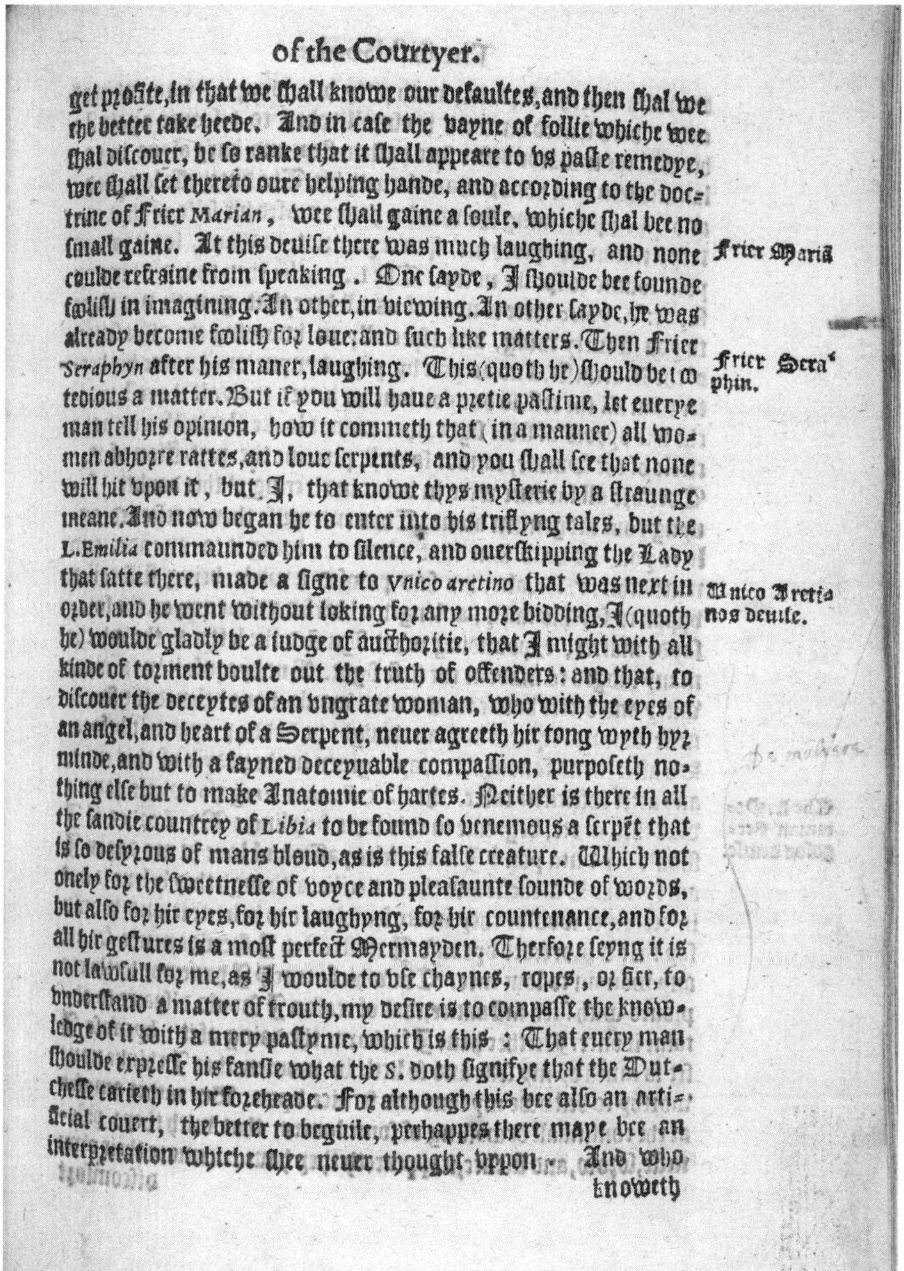

of the Courtyer.

get profite, in that we shall knowe our defaultes, and then shal we the better take heede. And in case the vayne of follie whiche wee shal discouer, be so ranke that it shall appeare to vs paste remedye, wee shall set thereto oure helping hande, and according to the doctrine of Frier Marian, wee shall gaine a soule, whiche shal bee no small gaine. At this deuise there was much laughing, and none coulde refraine from speaking. One sayde, I shoulde bee founde foolish in imagining. Another, in viewing. An other sayde, he was already become foolish for loue: and such like matters. Then Frier Seraphyn after his maner, laughing. This (quoth he) should be too tedious a matter. But if you will haue a pretie pastime, let euerye man tell his opinion, how it commeth that (in a manner) all women abhorre rattes, and loue serpents, and you shall see that none will hit vpon it, but I, that knowe thys mysterie by a straunge meane. And now began he to enter into his triflyng tales, but the L. Emilia commaunded him to silence, and ouerskipping the Lady that satte there, made a signe to Vnico aretino that was next in order, and he went without loking for any more bidding, I (quoth he) woulde gladly be a iudge of auctoritie, that I might with all kinde of torment boulte out the truth of offenders: and that, to discouer the deceytes of an vngrate woman, who with the eyes of an angel, and heart of a Serpent, neuer agreeth hir tong wyth hyr minde, and with a fayned deceyuable compassion, purposeth nothing else but to make Anatomie of hartes. Neither is there in all the sandie countrey of Libia to be found so venemous a serpẽt that is so desyrous of mans bloud, as is this false creature. Which not onely for the sweetnesse of voyce and pleasaunte sounde of wordes, but also for hir eyes, for hir laughyng, for hir countenance, and for all hir gestures is a most perfect Mermayden. Therfore seyng it is not lawfull for me, as I woulde to vse chaynes, ropes, or fier, to vnderstand a matter of trouth, my desire is to compasse the knowledge of it with a mery pastyme, which is this: That euery man shoulde expresse his fansie what the S. doth signifye that the Dutchesse carieth in hir forehead. For although this bee also an artificial couert, the better to beguile, perhappes there maye bee an interpretation whiche shee neuer thought vppon. And who

knoweth

Frier Mariã

Frier Seraphin.

Vnico Aretinos deuise.

FIGURE 10. Hair, shive, and handwriting in Baldassare Castiglione's *The Covrtyer of Covnt Baldessar Castilio, Devided into foure Bookes*, trans. Thomas Hoby (1577). © The British Library Board. Shelfmark 8404.ccc.25, sig. C5v–C6r.

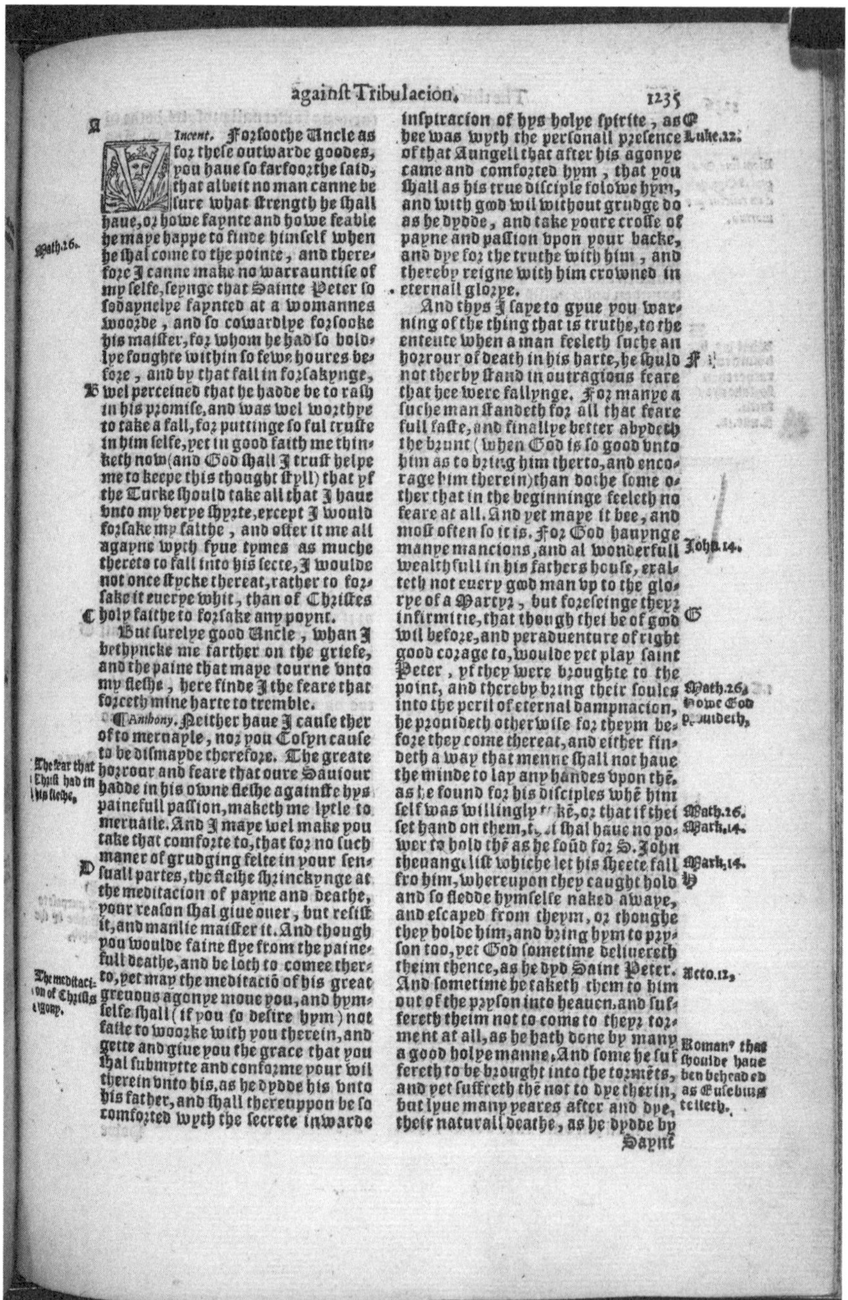

against Tribulacion. 1235

A Vincent. Forsoothe Uncle as for these outwarde goodes, you haue so farfoorthe said, that albeit no man canne be sure what strength he shall haue, or howe faynte and howe feable he maye happe to finde himself when he shal come to the pointe, and therefore I canne make no warrauntise of my selfe, seynge that Sainte Peter so sodaynelye faynted at a womannes woorde, and so cowardlye forsooke his maister, for whom he had so boldlye foughte within so fewe houres before, and by that fall in forsakynge,

Math.26.

B wel perceiued that he hadde be to rash in his promise, and was wel worthye to take a fall, for puttinge so ful truste in him selfe, yet in good faith me thinketh now (and God shall I trust helpe me to keepe this thought styll) that yf the Turke should take all that I haue unto my verye shyrte, except I would forsake my faithe, and offer it me all agayne wyth fyue tymes as muche thereto to fall into his secte, I woulde not once stycke thereat, rather to forsake it euerye whit, than of Christes

C holy faithe to forsake any poynt.

But surelye good Uncle, whan I bethyncke me farther on the griefe, and the paine that maye tourne unto my fleshe, here finde I the feare that forceth mine harte to tremble.

Anthony. Neither haue I cause ther of to meruayle, nor you Cosyn cause to be dismayde therefore. The greate horrour and feare that oure Sauiour hadde in his owne fleshe againste hys painefull passion, maketh me lytle to meruaile. And I maye wel make you take that comforte to, that for no such maner of grudging felte in your sen-

The fear that Christ had in his fleshe.

D suall partes, the fleshe shrinckynge at the meditacion of payne and deathe, your reason shal giue ouer, but resist it, and manlie maister it. And though you woulde faine flye from the painefull deathe, and be loth to comee thereto, yet may the meditació of his great greuous agonye moue you, and hymselfe shall (if you so desire hym) not faile to woorke with you therein, and gette and giue you the grace that you shal submytte and conforme your wil therein unto his, as he dydde his unto his father, and shall thereuppon be so comforted wyth the secrete inwarde

The meditacion of Christs agony.

inspiracion of hys holye spirite, as

E hee was wyth the personall presence of that Aungell that after his agonye came and comforted hym, that you shall as his true disciple folowe hym, and with good wil without grudge do as he dydde, and take youre crosse of payne and passion upon your backe, and dye for the truthe with him, and thereby reigne with him crowned in eternall glorye.

Luke.22.

And thys I saye to gyue you warning of the thing that is truthe, to the enteute when a man feeleth suche an horrour of death in his harte, he shuld

F not therby stand in outragious feare that hee were fallynge. For manye a suche man standeth for all that feare full faste, and finallye better abydeth the brunt (when God is so good unto him as to bring him therto, and encorage him therein) than dothe some other that in the beginninge feeleth no feare at all. And yet maye it bee, and most often so it is. For God hauynge manye mancions, and al wonderfull wealthfull in his fathers house, exalteth not euery good man up to the glorye of a Martyr, but foreseinge theyr

John.14.

G infirmitie, that though thei be of good wil before, and peraduenture of right good corage to, woulde yet play saint Peter, yf they were broughte to the point, and thereby bring their soules into the peril of eternal dampnacion, he prouideth otherwise for theym before they come thereat, and either findeth a way that menne shall not haue the minde to lay any handes upon thē, as he found for his disciples whē him self was willingly takē, or that if thei set hand on them, they shal haue no power to hold thē as he foūd for S. John theuangelist whiche let his sheete fall

Math.26. Howe God prouideth.

Math.26. Mark.14.

Mark.14.

H fro him, whereupon they caught hold and so fledde hymselfe naked awaye, and escaped from theym, or thoughe they holde him, and bring hym to pryson too, yet God sometime delivereth theim thence, as he dyd Saint Peter. And sometime he taketh them to him out of the pryson into heauen, and suffereth theim not to come to theyr torment at all, as he hath done by many a good holye manne. And some he suffereth to be brought into the tormēts, and yet suffreth thē not to dye therin, but lyue many yeares after and dye their naturall deathe, as he dydde by Saynt

Acto.12.

Roman' that shoulde haue ben beheaded as Eusebius telleth.

FIGURE 11. Flax shive in margin of *The Works of Sir Thomas More Knyght [. . .] Wrytten by Him in the Englysh Tonge* (1557) © The British Library Board. Shelfmark C.11.b.15, sig. 2Ii6r.

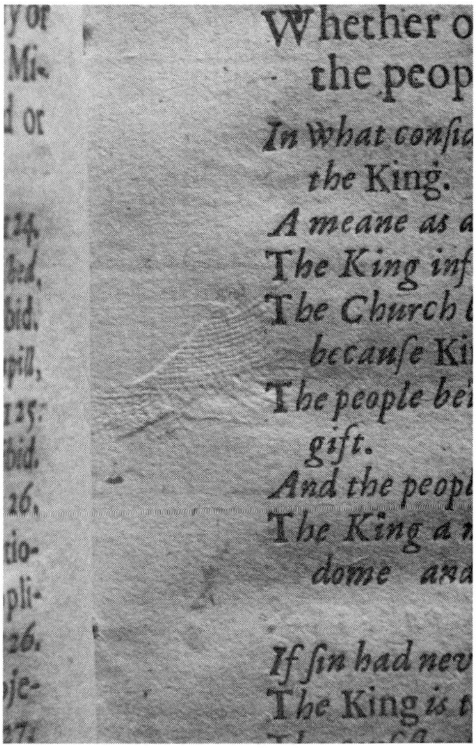

FIGURES 12 AND 13. Cloth swatch embedded in paper in "The Table of the Contents of the Book" of Samuel Rutherford's *Lex, Rex; The Law and the Prince* (1644). Fisher Library, University of Toronto, Forbes 00804, sig. C2r. Photographed with backlight and with raking light by Randall McLeod. Images courtesy of the Thomas Fisher Rare Book Library.

early modern writers and readers as nonhumans acting in humanistic ways. It is as impossible for me as it was for Hooke not to see some scatological humor in the textual movements of the bookworm, to imagine Rabelaisian bookworms intentionally defecating their way through pages of poulter's measure.

Rather than abandoning such playfulness, I see the value of grounding it such that the nonhuman ecology of the book and the human ecology of the book (including human disposition to narrativize, instrumentalize, and humanize) can coexist. The bookworms Hooke imagines annotate and digest books in different kinds of ways. John Carter and Nicolas Barker tell us that the bookworm "track is occasionally useful in detecting made-up copies" (copies where one or more missing leaves of one copy have been substituted with leaves from a different copy).[49] The holes that bookworms leave behind also offer detailed information about the nature of the page and, more specifically, the place where a book was read and/or stored. As biologist S. Blair Hedges has demonstrated, wormholes "document the detailed biogeographic history of two putative species of invasive wood-boring beetles."[50] (Technically, bookworms are beetles—or, rather, beetle larvae worming their way out of the book, eating their way toward a winged metamorphosis.) The two species of what we call bookworms did not coexist in the same book or even overlap geographically: the common furniture beetle (*Anobium punctatum*) was native only to northern Europe until the end of the nineteenth century, while the Mediterranean furniture beetle (*Oligomerus ptilinoides*) was native only to southern Europe during that time. Peter Stallybrass jokes about the religious significance of these different species: the "Protestant bookworms" found north of Paris chew more direct lines of escape, while the "Catholic bookworms" found south of Paris meander and often leave more ovate exit holes.[51] From the point of view of a wood-boring beetle or bookworm, book pages (and woodblock prints and bindings, etc.) were incubators for eggs. Upon maturing, adult beetles emerge, leave the exit holes we see in books, reproduce, and die. But their exit strategies and the differing sizes of the exit holes leave behind what Hedges refers to as "trace fossils" that can reveal much about the history of a book, including where and when it originated and where it was moved through time.[52] Hooke may have been wrong about "husks of Hemp and Flax" being a bookworm's preferred source of nutrition, but the presence of bookworm holes alongside these husks of organic material reminds us, across a multicentury gap, of the dynamic nature of the book.

In modern papermaking terminology, bits of visible matter in paper are referred to as "inclusives." They are, most often, intentionally added to the vat to create an aesthetic effect. One can buy paper with embedded flower petals or even greeting cards with flower seed inclusives that will grow if you bury and water the page. Crane and Company, which has been making paper for the

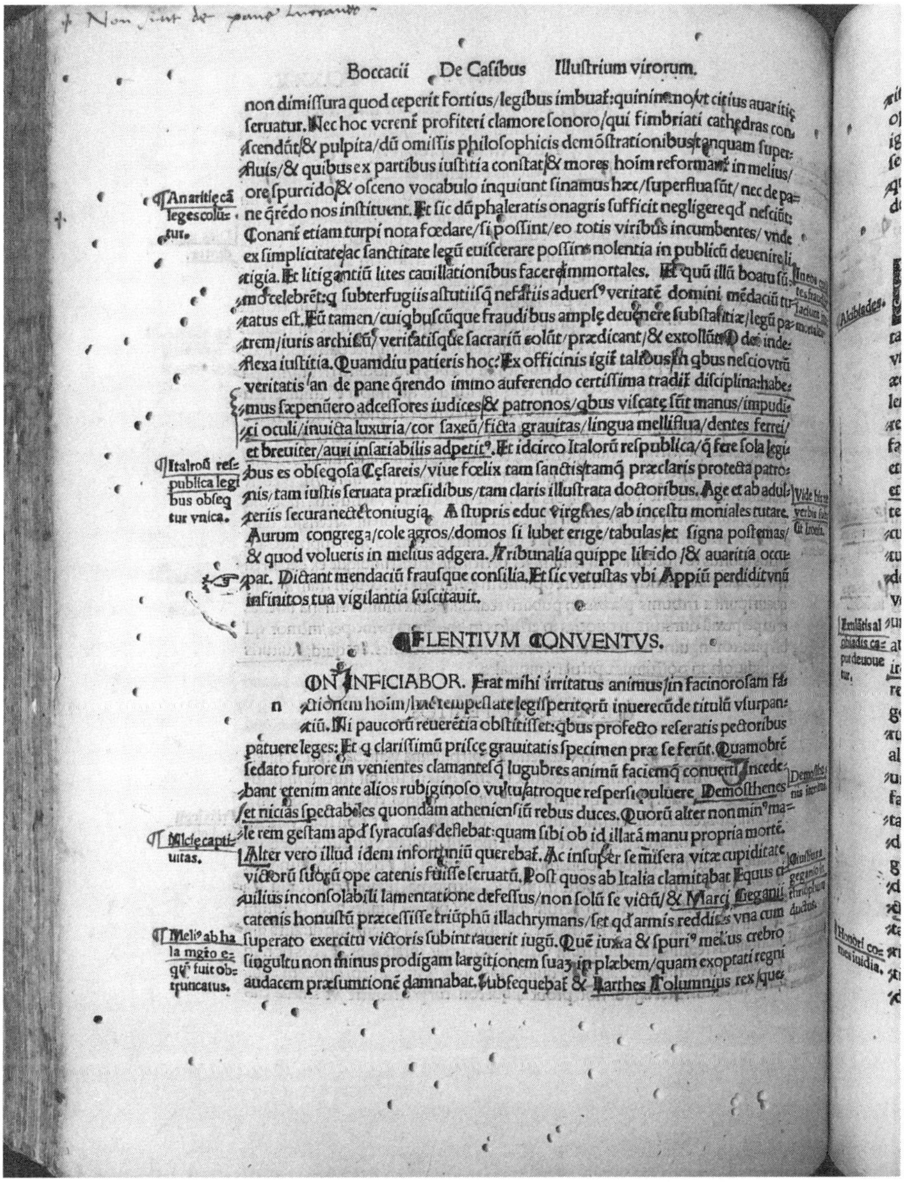

FIGURES 14 AND 15. Page and detail from book-wormed copy of Giovanni Boccaccio, *De casibas illustrium* ([1520]). Courtesy of the Department of Special Collections, Memorial Library, University of Wisconsin–Madison.

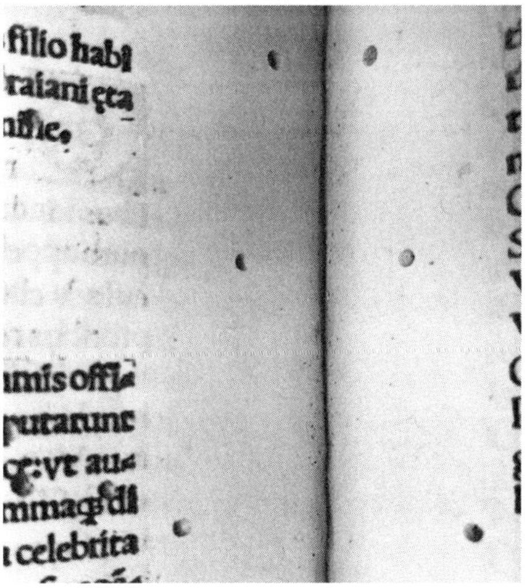

United States Treasury since 1879, uses inclusives such as security fibers and threads as "paper-based counterfeit deterrents."[53] Inclusives, whether intentionally added or not, have always been part of hand papermaking. As I have noted, they are in the rivers that supply water, they are mixed with the fibers that make up the pulp, they are on the felts used to separate and press the papers as they dry. Glitter can and is sometimes used as an inclusive in modern handmade paper, but if you walk into Robert Possehl's papermaking classroom, you will see one commandment written on the board: "No Glitter." The problem is that glitter never really goes away; three years after it is supposedly cleaned up, someone will find a few shimmering specks in a freshly pulled sheet of paper. It seems to me an apt analogy to early handmade paper: papermakers did their best to make uniform, homogenous, white paper from pure fibers and clean water, but the nature of—which is to say, the nature *in*—the page always manifests. With very little effort, we can find evidence in archival libraries and even in volumes digitized at medium resolution or better that will begin to reveal the social ecologies in which paper is produced and used.

Flax and the Materials of Memory

In their seminal study of book history, in a statement that this book takes up and expands, Lucien Febvre and Henri-Jean Martin claim that the growth of papermaking "was always conditional on the supply of its raw materials."[54] For instance, as papermaking spread westward from China, the technology had to accommodate geographical variations in native plant species. In Europe, it was "the extended cultivation of flax and hemp at the close of the Middle Ages, and the substitution of linen for wool in the making of underwear" that created a timely "rag base" for the emerging paper industry.[55] Furthermore, the symbiosis of the industrial and the natural did not necessarily end with a printed page. A book's paper might have been used as toilet paper, reused as fertilizer for flax plants, which were then harvested and converted into linen underwear before once again providing pulp for paper.[56]

All of these uses of flax—as plant, clothing, paper, and waste paper—remain simultaneously available. Present function does not obliterate past form. Clothing materials function economically and mnemonically, as Ann Rosalind Jones and Peter Stallybrass have so convincingly argued in *Renaissance Clothing and the Materials of Memory*.[57] John Donne repeatedly refers to his poems and prose as rags, summoning paper's previous form at moments of artistic frustration (performative or actual) with an insufficient medium. For instance, in a poem appended to *An Anatomy of the World* he writes: "Can these memorials, ragges of paper, giue / Life to that name, by which name they must liue?"[58] (Unlike the rest of the lines in the volume, the lines of "Elegie" are visually "ragged": every other line is indented.) Similarly, Donne refers to more than one of his letters as a "ragge of paper." With a letter to Sir Henry Goodyer, he includes a "ragge of verses."[59] Elsewhere in his letter-writing, Donne further puns upon "ragge" by referring to the "raggednesse" of his handwriting, and even allows paper rags and ragged writing to blur, as when he writes, "Sir, you are used to my hand, and, I think have leisure to spend some time in picking out sense, in ragges."[60]

Rags, flax that has been further processed through inhabitation, frequently slide into associations with paper before they are actually put to that use, as when Dogrel in Abraham Cowley's *The Guardian* says of the character Cutter, "Those breeches he wears, and his hat, I gave him: till then, he went like a Paper-mill all in rags."[61] Thomas Dekker and George Wilkins describe a writer so filled with vitriol that his words should appear on "paper made of the filthy linnen rags that had beene wrapt about the infected and vlcerous bodyes of beggers, that had dyed in a ditch of the pestilence."[62] While Dekker and Wilkins's example *imaginatively* invokes pestilence-infected rags, Jean Baptiste Van Helmont's works, translated into English in 1664, consider an actual dispute about pestilence by referring back to the flax origins from which paper is made. He considers "whether a Letter that is closed with a linnen thred, be a partaker of contagion" and claims "truly, paper is no lesse capable of contagion, than flax, from whence it is made."[63] Perhaps the most compelling literary example of a garment's persistent material associations appears in John Tatham's play *The Scots Figgaries: or, A Knot of Knaves*, where the Parson says, "Let *Iezabell* be brought before the Elders . . . let her smock be given to the Rag-men, it may come to be Paper, and her Condemnation writ in't."[64] Here the broad industries of rag recycling, papermaking, and textual production are collapsed into a miniature economy where the clothing of an accused woman is made into a text that condemns the very same woman.

This final example of the simultaneous significances of flax-made objects also calls up the association of flax spinning with women's labor. Juan Luis Vives pairs flax spinning and reading in the early education of a Christian woman.[65] Margaret Cavendish likens flax spinning and writing.[66] But, as Fiona McNeill has shown, because spinning did not provide a viable means of making a living in Renaissance England, the term "flax-wench" came to be associated with prostitution, and "spinsters and their tools routinely implied the opposite of chastity."[67] For instance, according to McNeill, in *Twelfth Night*, "when Sir Toby assures Sir Andrew that his hair 'hangs like flax on a distaff; and I hope to see a housewife take thee between her legs, and spin it off' (1.3.99), he associates spinning flax, an even lower-paid labor than spinning wool, with syphilis, which stripped the hair."[68] In general, female readers were more experienced with flax than male readers, and insomuch as the works I have cited suggest the legibility and aesthetic significance of paper grades, it is reasonable to infer that women would have possessed what we might call advanced literacy in the reading of textual media in Renaissance England. And not only in Renaissance England. As I discussed in Chapter 1, it was not until the second half of the nineteenth century that wood pulp became a viable alternative to the fibers recycled from rags; until then, writers and readers actively participated in the network of papermaking via flax cultivation and rag recycling.

Flax Speaks

In a bookish ecological debate, the prosopopoeic plant and animal interlocutors in *Lusus Serius: or, Serious Passe-time. A Philosophicall Discourse concerning the Superiority of Creatures under Man* offer a unique historical perspective on the nature of books. Printed in Latin in 1616 and in English in 1654, *Lusus Serius,* by the German alchemist Michael Maier, presents eight representative animals, vegetables, and minerals arguing for the title of "The King of All Worldly Things Being Under the Command of Man."[69] Social status, not natural prowess, is the subject of the debate among the plants, animals, and minerals, so rather than a lion arguing for his role as "King of the Beasts," a sheep and a calf argue on behalf of the four-footed beasts. A goose argues for the birds, an oyster for the fish, a bee for the insects, a silkworm for the creeping things, flax for the vegetables, and mercury for the minerals. Six out of the eight cite their textual roles as claims to preeminence: Sheep, Calf, Bee (wax tablets), and Flax all point out their usefulness in producing writing surfaces; Goose cites his quills; and Mercury cites his role in making movable metal type.

In *Lusus Serius,* Flax tells his own story as well as any history of the book has. Flax maintains that his growth "requires abundance of Labor in the preparation," but that humans are willing to go to all the trouble because of the benefits reaped from Flax.[70] He then describes the process of planting and manuring the seeds, of tending the plant, and of tending the harvest in preparation for making medicine, oil, or linen.[71] The actual making of linen is especially difficult, and Flax offers a detailed account of cultivating, harvesting, combing out the seeds, allowing the plants to putrify in water, drying, cooking, pounding, separating with cleavers, separating with hackles, spinning, washing, drying (again), bundling, weaving, and finally cutting and stitching.[72] The time lapse between all these steps from flax sowing to linen wearing was more than a year.[73]

Even after this tedious, time-consuming process of working flax, there are still remaining bits of what Flax calls the "rind"—which appear in clothing and later in paper as shives.[74] Converting flax to linen was worth all the trouble because linen had a double life; after praising the uses of linen "while it is whole," Flax speaks of the value of linen rags:

> When Linnen hath been so oft foul'd and apply'd to so many uses, and that so long, till it fall in pieces, behold a new production. For the raggs both of Linnen and cloth, be they never so much sullied, or torne, are gathered together to be carryed to the Paper-mill, where they make their choice and wash them, and then steep them in water till they be in a manner putrified, then they beat them into a kinde of poultice, which being made very thin, and besprinkled with water, or laid upon little nets of Wire, is made Paper . . . [which] brings innumerable helps to the conversation of man.[75]

Even here, Flax emphasizes his role in human "conversation." Flax concludes that "from *Flaxe* and Linnen there arises a new thing . . . like the Phænix out of her owne ashes."[76]

In Flax's closing arguments he stakes his reputation on his role in the history of the book. Flax argues that printing on paper is superior to writing on animal skins because it makes possible the mass dissemination of ancient writings recorded on such older, scarcer forms of media.[77] He points out the cost of materials for writing and reading and claims that "we must necessarily inferre, that in that time men of lower Fortunes were withdrawn and deterr'd from their Study, by those two great difficulties, the matter upon which they should write, and the dearnesse of Bookes."[78] He gives examples of poor philosophers writing in the sand for lack of materials and builds to the conclusion that, "by the Courtesie of *Flaxe*, the poorest man may have enough wherewithall to read, whereby to learne, and wherein he may dispose all things he hath either read, or hath been taught."[79] Flax's material cost-benefit analysis of the textual dissemination of ideas might as well be a pitch for the Reformation Bible or *The Norton Anthology.*

The grandest claim that Flax makes to preeminence, though, is that he is the fundamental matter not only of paper, but also of the entire book:

> Pray look upon printed books, of what consist they? marry *Flaxe*, and for the Letters the greatest part of them is Oyle of Linseed thickned, unto which I adde, some Paper burn'd till it be black, (as it well may be) or any other smoak made of Linseed, you have an Inke fit to print any

> Book. Therefore a whole printed book consists onely of *Flaxe*. Nor can any other claime a share in it, whether you will consider the Threads, or any thing else by which it is bound.[80]

Flax argues that in a book bound only by stitching, he is solely responsible for all of the materials (a significant claim for flax that I have not found elsewhere). Mercury, however, is unimpressed. Speaking on behalf of the minerals, he undermines Flax's claim with an argument that is surprisingly proto-industrial: he (Mercury) makes metals, and metals make machines, and machines make the rest of nature serve as fodder. Mercury's metal printing type is superior to all the other writing materials, he claims, so long as Flax's paper and printing ink will serve as his "drudges."[81]

Vaughan's Palimpsested Pages

If Mercury is superior in *Lusus Serius*, he is noticeably absent from Vaughan's poem, which focuses on the ecological remainders that a reader could have identified, not on the technological apparatuses used by printers. When Vaughan looks at his Bible's pages, he sees seeds and flax and plants and linen and people. These things, like the words printed on them, have meaning for him. Like many seventeenth-century readers, he still lives in close proximity to the materials that make his paper. He inherits "1 small flax wheele," used for spinning flax into thread, at his father's death.[82] He wears clothing that can be recycled when it grows threadbare. And he knows the value of various types of paper. Had he happened upon a copy of *Lusus Serius* in the year he was completing the second edition of *Silex Scintillans*, Vaughan might have been pleased to see two of his own works advertised at the back of the book on a supplemental quire of paper visibly inferior to the paper on which the primary text was printed.[83]

Unlike *Lusus Serius* and *Silex Scintillans,* Vaughan's Bible is not extant. The plants and animals "scatter'd thus" in its bookish form have probably been further scattered. Perhaps bookworms fed on the husks of flax in his "aged book." We can study the "flagrantly polychronic" Archimedes palimpsest described by Jonathan Gil Harris, on which "medieval liturgical script, modern forged image, and classical philosophical text" are layered and still discretely legible, but we cannot examine Vaughan's Bible and study its polychronic layers.[84] Vaughan does give us some clues about the Bible, though. We know that he, or the poetic persona of *Silex Scintillans,* learned to read from his Bible, a "cheap Book" that was given to him as a child.[85] From Vaughan's scripture quotations, we are fairly certain his Bible was an Authorized Version (first printed in 1611), with wording heavily influenced by Tyndale's translation. Like many extant Bibles from which children (and adults) learned to read, Vaughan's Bible probably had markings in it—doodles, pen trials, practice letters, and maybe the death- and birth-records of those whose lives intersected it.

We know Vaughan knew his Bible's ecological origins. (Like many scholars, I read "The Book" as a poem about a Bible.[86]) He observes that his Bible's paper is made from flax and that its cover is made from the wood of a tree and the skin of a "beast" (probably a calf). "To see the world of nature *as* a book was a Medieval and Renaissance commonplace," writes Alan Rudrum; "Vaughan's originality is to see the world of nature *in* a book."[87] By itemizing the ecological composition of a Bible, Vaughan draws a reader's attention beyond—and at times away from—transcendent religious ideas and toward its mundane matter. Discussing "The Book" as a prime example of "hylozoic poetry," Diane Kelsey McColley claims, "Part of ecological thinking is knowing where our artifacts come from, with what cost to the earth, to habitats, to species, to individuals of those species, and in human labor."[88] "The Book" itself, printed on flax-flecked paper like the Bible it describes, similarly recalls the natural history of the media through which it is conceived.

While the poem emphasizes textual ecology as such, it further insists upon the poetic potential of flora and fauna visible in sixteenth-and seventeenth-century texts. The natural media are part of the message. "The Book" suggests that the visible traces of textualized animals and plants could inflect the acts of reading and interpretation. Composed and printed during a century of religious reformation and scientific revolution, the poem raises a crucial question: How should one resolve the tension between ideas and the natural matter that mediates those ideas in the forms of texts? For Vaughan, specific questions about the roles of plants and animals in a Bible can lead one to meditate on materialism, religion, eternity, corruption, and mediation. Harris claims that "English Renaissance writers repeatedly recognize the polychronic dimensions of matter—the many shaping hands, artisanal and textual, that introduce into it multiple traces of different times, rendering the supposedly singular thing plural, both physically and temporally."[89] "The Book" is exceptional for its ability to draw multiple temporalities and actors (human and nonhuman) into a conversation about the translation of ideas to matter that is centered on concerns about the textual corruption of Bibles.

The structural features of the poem place special emphasis on lines 20–21, "Which makes me wisely weep and look / On my own dust," a temporal moment of connection among natural things, an "aged book," an "Eternal God," and a reader who has become acutely aware of his own mortality. Lines 20–21 gather all the scattered, corruptible textual elements into an intensely personal and poetic insight. It is the moment when the forms of Vaughan's Bible effect meaning, to use McKenzie's language.[90] In the midst of the fourth stanza, there is a decided shift as the poet turns his gaze away from the animal skin that covers the book and toward his own skin.[91] Line 20 acts as a *volta* in the poem, shifting the emphasis from book to reader and from description to meditation. Following the stanzaic structure to this point, line 20 should be the sixth and final line of the fourth stanza. However, it enjambs into the subsequent line, "On my own dust; meer dust it is," so that the expected sestet and quatrain form one ten-line stanza. (Even if the stanza were split into a sestet and quatrain, regularizing the structure of the poem overall, the stanzas would be enjambed rather than end-stopped, calling attention to lines 20–21.) Though the poem is lyrical, the speaker does not refer to himself until line 20, where the initial first-person pronoun ("me") appears. To this point in the poem, the speaker has referred to "they," "Thou," "this," and "it." Lines 20–21, where one expects the stanzaic break, are the most personal and intimate in the poem: the "me" and "my" in these two lines are the only first-person pronouns in the poem.

In the final stanza, the poet refers to himself again, but he does so in the third person ("him" in l. 29). By the time we reach the concluding lines, the charged moment of present-tense recognition, highlighted in lines 20–21, is past, and the poet concludes with a hopeful prayer: "Give him amongst thy works a place, / Who in them lov'd and sought thy face!" (ll. 29–30). The last stanza looks forward to another temporality: the day of resurrection, when God will "restore trees, beasts and men" and "make all new again" (ll. 26–27). These lines also introduce a paradox: at the day of restoration, when God restores all things to their perfect form, does the Bible remain a Bible, made from dead plants and animals? Or does it revert back to living matter? Are plants and animals more perfect than the printed Word of God? As one critic observes, "Other poets of the time liked to expatiate on the ultimate reunion of the body's dispersed dust," but "Vaughan gives an individual turn to the idea when he reflects that the flax . . . the tree . . . and the animal . . . will all be restored in eternity to their original form."[92] For Vaughan, the corruptible text is reanimated by the poet's awareness of the visible, persistent, perfectible materials from which it is made. The "them" is crucial because it is ambiguous; it allows one to imagine Vaughan seeking God's face in the words of plants-and-animals-as-Bible and in the things themselves, the plants and animals that are recalled by the literary form in which they survive.

Vaughan's Bible as a Book of Nature

The second stanza, in which Vaughan recalls paper's plant origins, best exemplifies the "ecology of texts," a term I use to put D. F. McKenzie's "sociology of texts" into conversation with the practice of making texts from natural resources.[93] McKenzie insists that bibliography is a study not only of "texts as recorded forms" but also of "production and reception," and he argues that bibliography and literary criticism can "show the human presence in any recorded text."[94] Bibliography and literary criticism can also recover nonhuman actors in the networks of textual production, and Vaughan's close reading of his own Bible models such critical work. Human and nonhuman presence is accounted for in Vaughan's consideration of the tree and the beast (the tree is "a *Cover* made" [l. 12], and the beast's skin is "spred" over the cover, or book boards [l. 18]), but it is in his consideration of his Bible's paper that Vaughan offers the most detailed consideration of the ways nature and society commingle in Renaissance texts. In six lines, Vaughan takes us from a material description of a seed that became paper to a metaphorical abstraction in which "*weeds*" signify ungodliness. Vaughan italicizes the key terms in this progression from *seed* to *grass* which is *drest* and *spun* into *linen* then *worn* (Vaughan uses "did *wear*" to fill out the line) by those whose lives were like *corn* or *weeds*. Nine—more than half of the words italicized in the poem—appear in this stanza. Most striking in this italic highlighting is the way in which Vaughan links the *papyr* of his Bible to its plant origins (further emphasized by the unusual spelling that recalls papyrus), then links the people who wore the *linen* with organic metaphors found in his Bible.[95] In Vaughan's reading, then, we see not only the vegetable origins of textual materials, but also the vegetable origins of metaphors inscribed on textual materials.

As a whole, this stanza draws together multiple temporalities into a specific, time-bound reflection on the matter palimpsested in a Bible's paper. It is crucial to recognize that though Vaughan does imagine the lives of linen-wearers he cannot see, Vaughan is not *simply* imagining a history for the paper. He is extrapolating from material proof. Just as he can knock on the cover of the Bible and hear the difference between book boards made from a tree's wood and those made from pasteboard, just as he can look at the cover of his Bible and see that it is made from an animal's skin, so he can look at the pages and see that they are made from flax. The evidence is in the paper, which is a palimpsest of natural growth, social processing, and textual production. Vaughan frames himself as a somewhat literate reader who can spell out the remains of plants and people in his pages, but he speaks to God as to an adept reader who perfectly comprehends the life of the plant and its wearers.

By invoking the lives of the linen-wearers, Vaughan also invokes what Jones and Stallybrass have called the "material mnemonics" of clothing: "The sixteenth and seventeenth centuries . . . are of particular interest in the history of clothing because clothes were still material mnemonics in metropolitan centers even as they were becoming the commodities upon which international capitalism was founded."[96] Rags hold economic value as papermaking materials, but rags also hold sentimental value: "The most worn-out piece of clothing can materialize an absent lover."[97] As I have shown, the idea that the past lives of clothes-wearers haunt the present life of paper is not original with Vaughan. However, Vaughan lets us see more levels of accretion in paper: the growth and harvesting of a plant, the handiwork of the spinners and clothmakers, and the lives of the wearers. Vaughan realizes that the new text has not obliterated the old memories woven and worn into its constituent materials.

James Simmonds offers a particularly valuable formalist reading that helps one see how flaxen object and poetic subject converse in "The Book." Focusing on themes of "decay and preservation, the continuous transfiguration of life to death and death to life" in the poem, Simmonds convincingly argues that Vaughan's interaction with nature was more bookish than experiential, and that his poetry is best read as conversant with seventeenth-century religion, not nineteenth-century Romantics.[98] In other words, "The Book" is a poem about ideas and matter, not a poem about being "green" or ecologically aware. Simmonds establishes the way "The Book" works within the meditative poetical form. He acknowledges that the poem "involves meditation on the creatures"—a type of meditative poetry that draws doctrine from natural observation—but claims that, in "The Book," the meditation on the creatures is part of a more "basic pattern of meditation on death, of which the book is an emblem" or *memento mori*.[99] "The subject of the meditation is death," he writes; "its object . . . is the Bible perceived as a material thing."[100]

One is tempted, in the spirit of material culture studies and book history, to fixate on the object, but Simmonds's statement that the subject is death reminds us that the poem, with all its ecological description, is rhetorical. I conclude by slightly amending Simmonds's insight: the subject of the poem is *corruptibility*, not *death*, and its object is a printed Bible made of culturally processed natural resources, a Bible that is a palimpsest of plants and animals, social circulation, religious tradition, and textual production. Nature is part of the book's poetic form. In a temporal moment of insight, Vaughan looks at the pages of his Bible and recognizes its polychronic layers. He imagines the past and sees the separate lives of actors. He looks at the present, the book in his hands, and sees the actors' deaths. The book he holds is dead matter: his Bible is corrupt. The reader is corrupt. His only source of hope is to look to the future and envision the restoration. Material corruption can be reversed, for the dead matter that inflects "The Book" when it appears in print also preserves the quintessence of that matter. Corruption and death in Vaughan's cosmos are not final. Matter is corruptible, but only to a point. The structure of "The Book" contrasts the earthly experience of material, time-bound creation in flux with the heavenly stability of a creator unbound by time or matter: "Eternal God," "Rock of ages," "knowing, glorious spirit."

For Vaughan, the interference—and inter-reference—of things with words is indicative of a broader tension between creator and created, between the supreme idea and the dusty matter of human experience. There is a sense of longing that threads *Silex Scintillans*, an urgent insistence on finding that which is lost. And Vaughan most often looks for lost truth in the Bible and in nature. "No book of seventeenth-century poetry bears more obvious signs of the impress of the Authorized Version of the Bible than *Silex Scintillans*," writes E. C. Pettet;[101] and yet, as another scholar has noted, "Vaughan is the only metaphysical poet ever to be accused of being a poet of nature."[102] The final couplet of "The Book" epitomizes the seeking poetry of *Silex Scintillans* that looks to the Bible and the Book of Nature for truth: "Give him amongst thy works a place, / Who in them lov'd and sought thy face!" (ll. 29–30). He seeks God's face in "them": the flora and fauna around him and the flora and fauna from which his Bible is fashioned. Vaughan's Bible and the Book of Nature merge in this closing line. Words are things, and both words and things, although—or possibly because—they are corruptible, allow the poet and his creator to converse in a temporal recognition that sees eternity in a seed of flax.

In the next chapter, which begins Part II, "Indistinct Ecologies," we shift away from recognition toward puzzlement, from the legibility of plant fibers to the muddle made by blotted ink on paper. Although Renaissance writers and readers understood the unique interactions between ink and paper, on the one hand, and ink and parchment, on the other, they do not seem to have fully registered how much the nonhuman materials of writing and revising influenced the metaphors they used to understand human narratives of achievement and failure. We turn, then, from Henry Vaughan to Shakespeare's English history plays, and especially to the story of Henry V as imagined in the Henriad (*Richard II*, *1* and *2 Henry IV*, and *Henry V*). As Vaughan does in "The Book," I examine a particular volume that, upon closer inspection, reveals a deeper narrative about its material and metaphorical ecologies.

PART II

INDISTINCT ECOLOGIES

CHAPTER 3

How to Read a Blot: Historiography and Renaissance Ecologies of Inscriptive Error

> When one uses antique media like stereoscopes, when one encounters unfamiliar protocols, like using a pay telephone abroad, or when media break down, like the Hubble Space Telescope, forgotten questions about whether and how media do the job can bubble to the surface.
>
> —Lisa Gitelman, *Always Already New*

In the gutter of one of the Folger Shakespeare Library's First Folios, in act 5, scene 2 of *The Second Part of King Henry the Fourth*, there is a large, 85 x 100 millimeter blot that recalls a Rorschach test.[1] Another reader might see something else entirely, but it looks to me like a large-headed person seated cross-legged with arms open toward the reader. One is tempted to read the blot onto the play and especially onto the scene it partially stains. When act 5, scene 2 opens, King Henry IV is pronounced dead and an uncomfortable bit of conversational avoidance ensues until Prince John addresses the awkwardness:

JOHN OF LANCASTER. We meet like men that had forgot to speak.
WARWICK. We do remember, but our argument
Is all too heavy to admit much talk.
JOHN OF LANCASTER. Well, peace be with him that hath made us heavy.
(5.2.23–26)

Prince Harry, now the soon-to-be-crowned King Henry V, is "him that hath made us heavy." Everyone in the room—the Earl of Warwick, the other three princes, and especially the Chief Justice—assumes that the "riot and dishonor" that "stain[ed]" the prince's youth will also stain the king's reign.[2] The group tries and fails to not talk about the blots that stain their new ruler. The actual stain on the page, the unusually large blot that is too obvious to overlook, almost begs to be read as savvy commentary.

Although the serendipitous juxtaposition of blot and scene may tempt one to read the blot onto the play, blots, be they accidental spills or intentional corrections, are textual defects that should be ignored, according to Francis Quarles. "If the Paper chance to have a Blot," Quarles writes in his 1640 book of maxims, "Remember, the Blot is no part of the Coppy."[3] Like Quarles, no respectable New Bibliographer would count this blot that appears in just one copy of the Folio as a textual variant worthy of notice. A blot like the one on Folio 47 is an accidental (and, it appears, also an accident).[4] Like the ecological accidentals, bits of organic matter embedded in the pages of handmade paper, discussed in the previous chapter, this blot is a nonessential feature of the text, one that does not represent the Folio's ideal form. "Remember, the Blot is no part of the Coppy." Quarles's methodology regarding blots, like the methodology of descriptive bibliographers, is attentive to accidentals—they matter as part of the physical manifestation or formal presentation of an individual text—but they do not truly count. The blot becomes, in Quarles's text and on the Folio page, something we read or process as something we do not read or process. Some marks on the page, then, are less significant than others. And yet accidental blots on the page have the curious effect of attracting a reader's eye, of emphatically calling attention to themselves.

What, then, should a literary reading do with a blot? Rejecting, with D. F. McKenzie, "a Platonic distinction between idea or essence on the one hand and its *de*forming, material embodiment on the other," we might yet retain a Socratic naivete, even an intentional critical humility, toward ink marks on a page and their relative significance.[5] Seeing the ecology of texts occasionally demands cultivated credulity. Rather than assuming that the material conditions of the text and/or that its uses have de-formed the essential text, one might ask how the mark we see on the page reveals something of the nature of the text as it has been transmitted to us. Focusing on the research itself and not on presumed outcomes, Rita Felski has argued, in the form of what she calls a manifesto, that instead of endlessly reusing the diagnostic, remedial model of literary criticism, "perhaps the time has come to resist the automatism of our own resistance, to risk alternate forms of aesthetic engagement."[6] Advocating a hybrid phenomenology or a neo-phenomenology, Felski calls for "an undogmatic openness to a spectrum of literary responses" and posits the value of "ways of thinking about aesthetic experience that do not hinge on the presumed superiority of literature or literariness."[7] Reading blots is one way to test the value of our critical patience with such open-ended engagements.

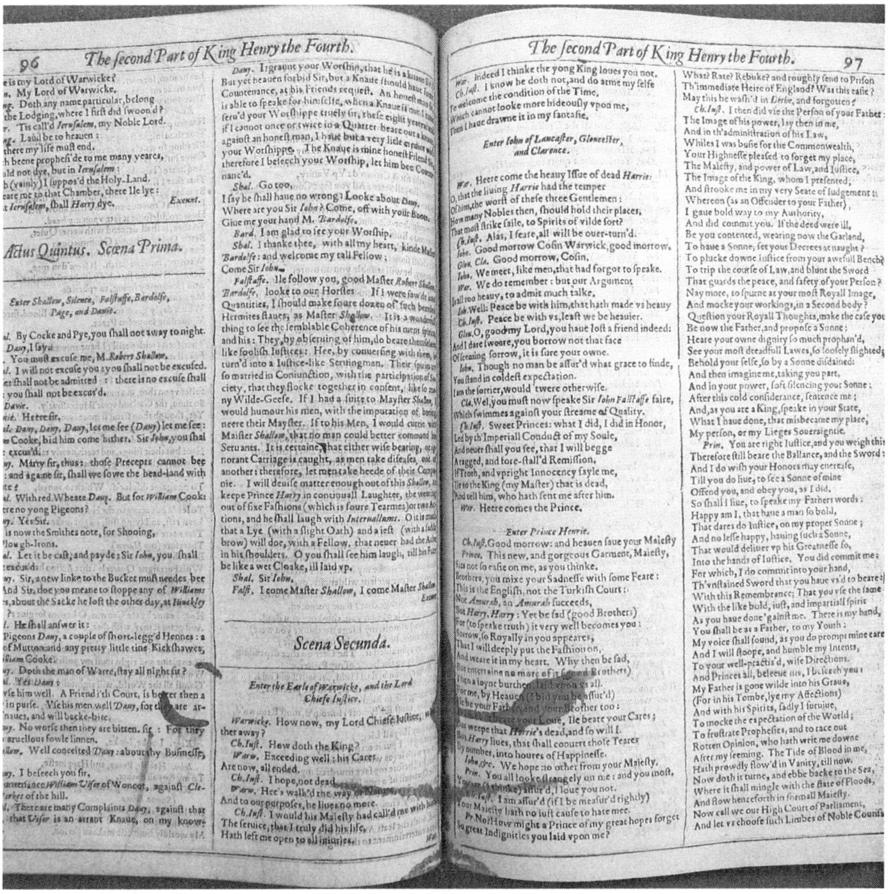

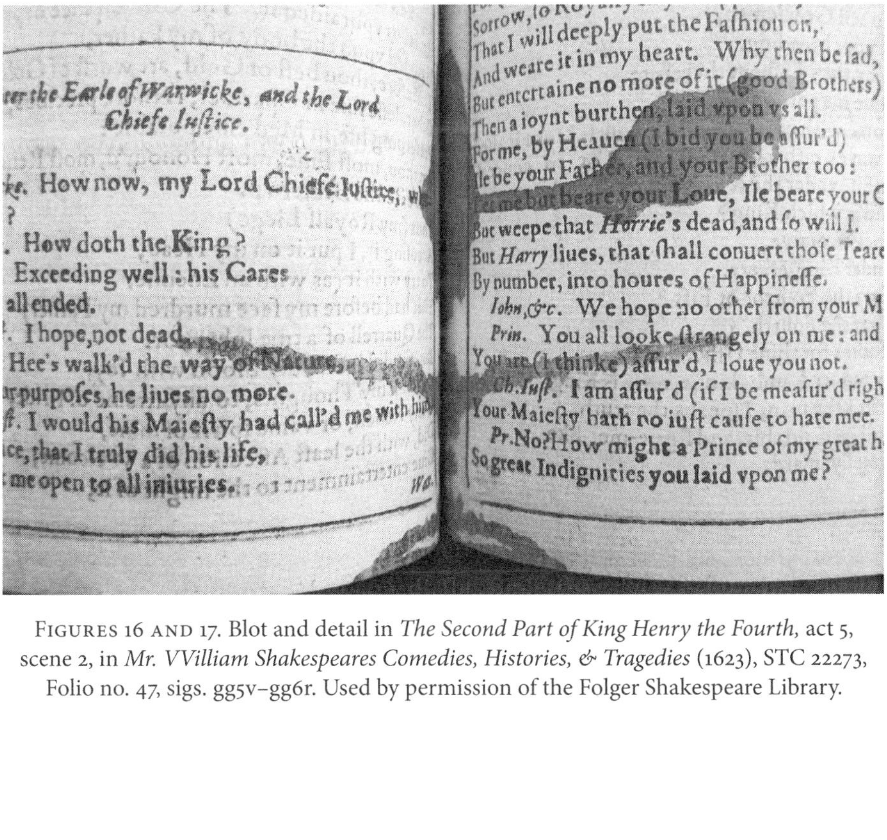

FIGURES 16 AND 17. Blot and detail in *The Second Part of King Henry the Fourth*, act 5, scene 2, in *Mr. VVilliam Shakespeares Comedies, Histories, & Tragedies* (1623), STC 22273, Folio no. 47, sigs. gg5v–gg6r. Used by permission of the Folger Shakespeare Library.

Unlike ecological accidentals embedded in paper on which a text is printed, stray blots and stains on a printed page are postproduction accidentals that indicate readerly interaction. Unintentional blots may tell us little about a historical reader's reaction to a text, but as I show in this chapter, they can tell us a great deal about how Renaissance writers and readers thought about error, correction, emendation, and, more broadly, forgiveness. This chapter demonstrates how to read a blot—and why to read a blot—by delving into the material interactions between ink and paper, but it also brings parchment into the conversation and focuses especially on the different kinds of corrections that each substrate allowed: in general, errors on paper were blotted, while errors on parchment were erased. We know that correcting a mistake on a computer and correcting a mistake on a page of paper printed from the same computer are wholly different actions with different vocabularies ("delete" or "cut" or "cut and paste" versus "cross out" or "x out" or "crumple up," for example) that leave different kinds of evidence of the changes that have been made. And it is the side-by-side comparison of these actions, vocabularies, and evidence that draws attention to differences, to textual (and cultural and social) practices grounded in textual affordances. Similarly, when we come to understand the side-by-side use of parchment and paper in Renaissance England, we begin to see how particular surfaces were used for particular tasks. But, as I will show, we also begin to see how textual affordances influence textual metaphors in subtle, but crucial ways that are not always obvious or legible to writers and readers.

Mixed Media, Mixed Metaphors

Among the troubled conversants in the opening lines of *2 Henry IV*, 5.2, the Chief Justice is especially anxious, even mordant:

> CHIEF JUSTICE. I would his Majesty [Henry IV] had called me with
> him.
> The service that I truly did his life
> Hath left me open to all injuries.
> WARWICK. Indeed, I think the young king [Henry V] loves you not.
> (5.2.7–10)

The service referred to by the Chief Justice recalls information supplied in a different play, the anonymous *The Famovs Victories of Henry the fifth.* In the play, the prince gives the Chief Justice "a boxe on the eare," and the Chief

Justice, taking seriously his charge to administer the king's laws, sends the prince to Fleet prison.[8] Knowing, as he supposes, the reckless prince's character, the Chief Justice assumes "all will be overturned," that the new king will replace virtue with vice, and he will get the worst of it (5.2.20). Prince John agrees: "Though no man be assured what grace to find, / You stand in coldest expectation" (5.2.31–32).

On Folio 47, one of the blot-figure's "arms" extends over part of this scene, just where Prince Harry enters on this anxious gathering (see Figures 16 and 17). The reach of the blot is indicated in the semidiplomatic transcription below with bold, underscored text:

> *Prin.* You all look**e strang**ely on me: and you most,
> Yo**u are (I thinke) assur**'d, I loue you not.
> ***Ch.Iust.*** I am assur'd (if I be measur'd rightly)
> **Y**our Maiesty hath no iust cause to hate mee.
> *Pr.* No? How might a Prince of my great hopes forget
> So great Indignities you laid vpon me? [column break]
> What? Rate? Rebuke? and roughly send to Prison
> Th'immediate Heire of England? Was this easie?
> May this be wash'd in *Lethe*, and forgotten?[9]

Shakespeare's Prince Harry, whose character has a theatrical habit of testing people in front of an audience (onstage and beyond the stage), is testing the Chief Justice—and maybe he is also playing one last trick before his metamorphosis. Unlike his previous tavern interrogations of Falstaff (after Gad's Hill in *1 Henry IV*, 2.4 and again with Poins in *2 Henry IV*, 2.4) or his eventual disguised wanderings around camp on the night before Agincourt (*Henry V*, 4.1), he does not discuss his plans beforehand, so we are not in on the trick. The scene plays out more like his interrogation of the traitors Scroop, Cambridge, and Grey in *Henry V*, 2.2, where he imitates the entrapping style of Nathan the Prophet in the interrogation of King David, a not insignificant connection to which I return at the end of this chapter.[10] It is a scene that revolves around the question of blame: was it just, Prince Harry asks the Chief Justice, to "rate, rebuke, and roughly send to prison / Th' immediate heir of England?" (5.2.71–72). But equally important is the question of forgiveness and forgetfulness: "May this be washed in Lethe and forgotten?" (5.2.73). The Chief Justice believes he was in the right, and he has already promised the other princes that he will not falsely beg "remission" for a sin he did not commit (5.2.39). He is true to his word, answering the charges without apology. In the end, he is not punished, but reinvested with the accoutrements of justice: the "balance" and the "unstainèd sword" (5.2.104, 115).

The prince's new outlook on an old offense signals a new person, as he tells his brothers the next column over, just beyond the reach of the blot:

And, princes all, believe me, I beseech you:
My father is gone wild into his grave,
For in his tomb lie my affections,
And with his spirits sadly I survive
To mock the expectation of the world,
To frustrate prophecies, and to raze out
Rotten opinion, who hath writ me down
After my seeming.
(5.2.123–30)

Although King Henry IV has complained that "riot and dishonor stain the brow / Of my young Harry" (*1 Henry IV*, 1.1.84–85), Prince Harry refigures his life as a palimpsest from which he can erase one version of his life story and replace it with a new story.[11] Likening his alteration from roguish prince to noble king as an act of removal and rewriting, Harry dismisses the record of his past sins as an erasable inscriptive error.

The clean narrative that Prince Harry envisions for his future as king is altered in not-so-subtle and important ways from the reformation narrative that he imagined as prince in *1 Henry IV*. In that account, famously spoken in poetry at the end of act 1, scene 2, the bad remains legible next to the good because Harry's reformation narrative plays upon the contrast between bright and dark:

So when this loose behavior I throw off
And pay the debt I never promisèd,
By how much better than my word I am,
By so much shall I falsify men's hopes;
And, like bright metal on a sullen ground,
My reformation, glitt'ring o'er my fault,
Shall show more goodly and attract more eyes
Than that which hath no foil to set it off.
I'll so offend to make offense a skill,
Redeeming time when men think least I will.
(*1 Henry IV*, 1.2.215–24)

1 Henry IV, then, relies on the visibility of error (bad as a "foil" to good), while *2 Henry IV* relies on the invisibility of error—on its erasure. The prince, it appears, is trying out various narratives, playing with historiography or "ways of writing history, of conceptualizing its methodologies and protocols."[12] Of the speech in which Prince Harry promises to "race out / Rotten Opinion," H. Austin Whitver writes, "With this speech [Prince Harry] can usher in an entirely new way to engage with the past: a historiography that both preserves the past and shapes it into narrative through the force of monuments."[13] We might observe that Harry is playing not only with ways of writing history, but also with the means of writing history: monuments, yes, but especially paper and parchment.

The placement of the blot in Folio 47 strikes a notable contrast on the page, spreading out as it does on this leaf where the characters reimagine and retell narratives that rely on the language of effacement such as "washed," "forgotten," "remission," "convert," "unstained," and "raze."[14] And it calls attention to the two primary methods that one might use to revise a written or printed record in Renaissance England: blotting and erasing. The razing or erasure of the historical record that Prince Harry imagines would be impossible on the pages of the Folio in which he appears. Razing or scraping can effectively remove ink from animal substrates—parchment or vellum—but not from the vegetal fibers of paper. Working with paper, a writer or reader will sometimes blot up ink that has not dried (removal), but more often corrects a mistake by covering ink with more ink, that is, by making errors hypervisible. On the material page, and by extension as a metaphor on stage, a blot can function as paralipsis, as an explicit gesture toward a thing that is emphasized even as it is overtly ignored. Remember, though: the blot is no part of the copy. It is a statistical anomaly, a meaning-less data point.

How then might the complexities of paper surfaces be brought into literary analysis? Randall McLeod offers a playful, engaging model in "Obliterature: Reading a Censored Text of Donne's 'To his mistress going to bed,'" where he demonstrates how one might re-literate that which is supposed to be ob-literate. McLeod begins by showing that the lines of one manuscript copy of Donne's famous poem do not appear as advertised by the editors of the variorum edition: "totally obliterated by ink smears."[15] Letters and even whole words can be seen with the naked eye "among the swirls of censoring ink": "'defye,' for example, at the end of the first line, defies 'totally obliterated.'"[16] Using infrared light to read further beneath the blot, McLeod is able to reveal "three unprecedented words," three unique textual variants recorded in the poem.[17] McLeod's obliterary approach exemplifies how one might transparently and creatively conjoin presentist and historicist interests in material poetry, and how such a transparent collapsing of subject and object in time enables insightful readings.[18] It also argues for the sort of open-ended curiosity that an ecology of texts requires: McLeod uncovered previously unknown textual variants for a famous poem, but only because he decided to read—and read through—a blot to see what he could find. As Mark Bland suggests, McLeod is "interested in the ways in which books and manuscripts contain signs that are other than the text, or that are not immediately visible to the eye," but the insights he gains allow him to "map patterns" and to show that "there is more than one possible 'reading' of the physical book."[19]

A more literary (and less bibliographical) model for reading blots is embedded in Wendy Wall's "Reading for the Blot: Textual Desire in Early Modern England." Attending to the recursive relationship between textual materials and literary metaphors, Wall explores the ways in which authors reconfigure symbolic relationships between people and the tools they use to record ideas on paper.[20] This material/metaphorical recursivity and reconfiguration grows even more intriguing and significant when we begin to recognize the ecological underpinnings of the textual materials that underpin literary metaphors. According to Jonathan Walker, "The problem of how to conceptualize the relationship between a literary text's semantic elements and the physical manifestation of those elements within a particular artifact is a challenge that textual critics have yet to lay to rest."[21] At bottom, Walker is taking up the question of what counts—what is allowed to count. Are we sure that the blot is no part of the copy? Advocating a "promiscuous practice of reading" that "cross[es] back and forth from literary and bibliographical fields and lexicons, and from semantic and nonsemantic features of the page," Walker finds literary significance "in the physical relationship between dramatic language and the peculiarities of the textual environment where that language is written or printed."[22]

Taking up Walker's invocation of "textual environment" and defining it as broadly as possible, this chapter explores the entangled relationship between language and environment, between the stories we want to shape about ourselves and others and the materials that shape our expression. There has been a tendency, in book history studies, to forget that books are objects made from nature. As I demonstrated in my focus on ship sails, rags, and flax in Part I of this book, the substrates used to record time or realms are not nearly as simple, self-evident, and inevitable as they sometimes seem in retrospect.[23] They are found objects made from nature that writers and readers in particular societies accept as—and, in the process, translate into—memorials. The materials, like the histories they record are subject to revision.

Fixing History: Metaphors of Writing and Revision in the Henriad

2 Henry IV has more incidence of the word "remember," more "letters," and more erasable writing "tables" than any of Shakespeare's other seven English history plays. It has more "paper," more "memory," and more "scrolls" than the other plays in the Henriad.[24] It has more "books" than the other three Henriad plays combined. But no "blots," less "sin" than other plays in the Henriad, and less "shame" and fewer "faults" than any of the other English history plays.[25] Other plays in the Henriad repeatedly refer to blots—*Richard II* has more blots than any other Shakespearean play—but *2 Henry IV* tends to emphasize a different kind of media interaction altogether. We might even say that the play is notably unblotted: *2 Henry IV* happens to be the only play in any of the English history plays that does not have at least one occurrence of the word "blot" (or some form of that word).[26] Robert N. Watson cites *Henry V* as "a history play obsessed with the difference between reality and its representations"; filled as it is with references to the mediation and erasure of memories, we could say that *2 Henry IV* is a play obsessed with how reality can be recorded and revised.[27] Rumor, misreport, misremembrance, and misrepresentation dominate *2 Henry IV*, and, as other scholars have demonstrated, the mixing of oral and written reports, of human servant pages and nonhuman paper pages, is one of the important means by which fake news undermines the project of history making. In this chapter, however, I want to allow the blotted page of *2 Henry IV* to guide our eyes back toward surfaces and thereby back toward the crucial role that mixed media—particularly mixed substrates—play in undermining the writing and revising of history. I argue here that the mixed media in the Henriad, and especially in *2 Henry IV*, inflect and allow the sort of mixed messages we see in the examples above. Media and messages are inseparable not only because mediation is a necessary condition of human communication, but also because the plays repeatedly draw attention to the various media that can be used to convey ideas.

Two quarto editions of *2 Henry IV* were printed before it appeared in the First Folio: a 1600 edition known as Quarto A and a reissue the same year—not truly a second edition—known as Quarto B. One blot, so to speak, on the Quarto versions is that they print profanity in the form of Jesus's name being taken in vain.[28] A reader of the Folger's copy of Quarto A seems to have had strong feelings about the profanity that she or he found there. If we are to believe the item note in the online catalogue, "the words 'Iesu' and 'Iesus' are *highlighted* in ink throughout the text" (emphasis mine).[29] Most likely, however, the reader of this text, who went through and carefully blotted out each instance where a character breaks the third commandment and invokes the Lord's name in vain, was trying to hide, not highlight, the profanity from future readers. Over time, it seems the reader's blots faded so that now the very thing meant to be censored is accentuated. In one place, for instance, Falstaff speaks the well-known lines "We haue heard the chimes at midnight M. Shallow." If the blotting had succeeded, we should have read Master Shallow's response as "[ ] the daies that wee haue seene." Instead, we find "Iesus the daies that wee haue seene."[30]

In all three of the earliest versions of *2 Henry IV*, even before act 1 begins, the character Rumor appears onstage to speak an induction that is equal parts humorous and horrifying. It is an intriguing opening that sets the tone for the whole play. *Henry V* has a more well-known, celebratory opening chorus ("O, for a muse of fire," lines that appear for the first time in the First Folio), but in *2 Henry IV*, the play that chronologically precedes it, the opening scene is set by a liar who, Iago-like, boasts about lying and then seems to let the audience in on the truth.[31] From the outset, then, the audience is mired in the very issues the play takes up in act 1, scene 1: how to sift out truth not just from lies, but also from half lies and hopeful lies and errors of transmission and exaggerated, dubious claims to factual information. Rumor confides in the audience that she is running out ahead and spreading the false report that Prince Hal was defeated by Hotspur at Shrewsbury:

> The posts come tiring on,
> And not a man of them brings other news
> Than they have learnt of me. From Rumor's tongues
> They bring smooth comforts false, worse than true wrongs.
> (Induction 37–40)[32]

These posts, as Rumor explains, are just arriving at the "worm-eaten hold of ragged stone, / Where Hotspur's father, old Northumberland, / Lies crafty-sick" (Induction 35–37).[33] As act 1, scene 1 opens, Rumor's induction is proven true: in Loren M. Blinde's words, "Narrative after unreliable narrative opens 1.1."[34] The 235 lines of this opening scene are filled with references to oral and written reports, and to the affordances of each kind of media source. In the scene we find false report by word of mouth, true report by word of mouth that is likened to a printed record, false report by word of mouth that rewrites oral history and written record, and an allusion to a writing table, an erasable note-taking device that might be thought to exist on the continuum between oral report and written record (see 1.1.231).[35] Northumberland first hears Bardolph's happy report that Hotspur has killed Prince Hal and taken Falstaff prisoner, then hears the

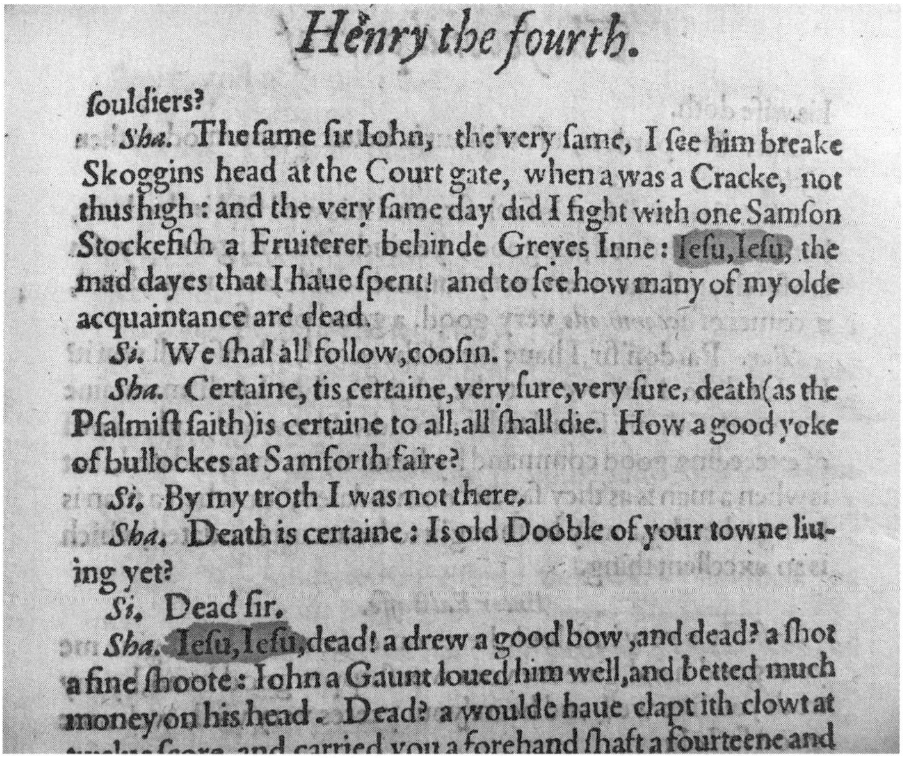

Henry the fourth.

ſouldiers?

Sha. The ſame ſir Iohn, the very ſame, I ſee him breake Skoggins head at the Court gate, when a was a Cracke, not thus high: and the very ſame day did I fight with one Samſon Stockefiſh a Fruiterer behinde Greyes Inne: Ieſu, Ieſu, the mad dayes that I haue ſpent! and to ſee how many of my olde acquaintance are dead.

Si. We ſhal all follow, cooſin.

Sha. Certaine, tis certaine, very ſure, very ſure, death (as the Pſalmiſt ſaith) is certaine to all, all ſhall die. How a good yoke of bullockes at Samforth faire?

Si. By my troth I was not there.

Sha. Death is certaine: Is old Dooble of your towne liuing yet?

Si. Dead ſir.

Sha. Ieſu, Ieſu, dead! a drew a good bow, and dead? a ſhot a fine ſhoote: Iohn a Gaunt loued him well, and betted much money on his head. Dead? a woulde haue clapt ith clowt at

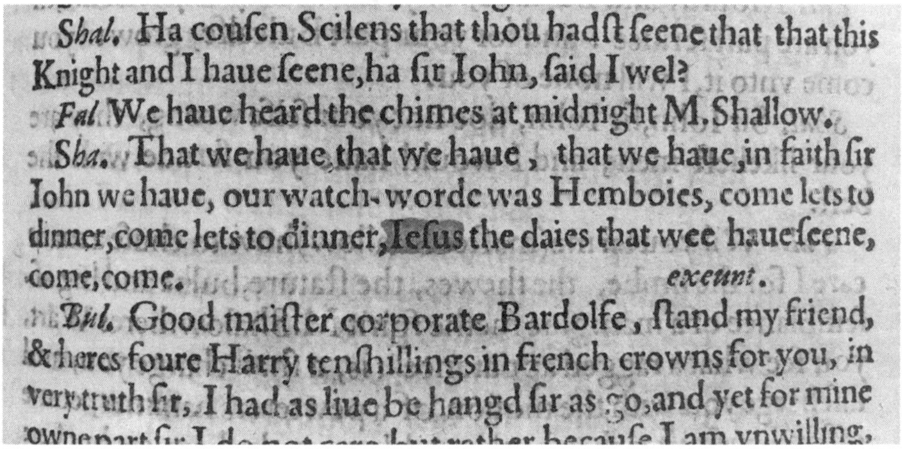

Shal. Ha couſen Scilens that thou hadſt ſeene that that this Knight and I haue ſeene, ha ſir Iohn, ſaid I wel?

Fal. We haue heard the chimes at midnight M. Shallow.

Sha. That we haue, that we haue, that we haue, in faith ſir Iohn we haue, our watch-worde was Hemboies, come lets to dinner, come lets to dinner, Ieſus the daies that wee haue ſeene, come, come. *exeunt*.

Bul. Good maiſter corporate Bardolfe, ſtand my friend, & heres foure Harry tenſhillings in french crowns for you, in very truth ſir, I had as liue be hangd ſir as go, and yet for mine owne part ſir I do not care but rather becauſe I am vnwilling,

Figures 18 and 19. "Highlighted" blots in *2 Henry IV*, in William Shakespeare, *The Second Part of Henrie the Fourth* (Quarto A; 1600), STC 22288, sigs. E4r and F2r. Used by permission of the Folger Shakespeare Library.

mixed report that Travers has gathered from two variant sources.[36] It is Morton, the eyewitness, who transmits the true report that Harry has slain Hotspur. Morton arrives, and, before he speaks, Northumberland likens his brow to a "title leaf" that "foretells the nature of a tragic volume" (1.1.70–71). Hotspur is dead. The scene ends with Northumberland arranging for messengers to be sent out to potential allies—but with an important difference. "Get posts and letters," says Northumberland (1.1.234). The posts who leave Northumberland's presence leave with written records.[37]

Despite, or perhaps because of, the play's preoccupation with recording, *2 Henry IV* is also repetitive. Reviewing the stage history of the Henriad, Lois Potter writes, "A number of directors have conflated the two parts of *Henry IV*, usually by cutting scenes in Part 2 that largely repeat those of Part 1 and by removing passages in which characters . . . recall the past."[38] Such a staging decision is understandable, and yet it is worth recognizing that nothing is ever perfectly or, one might say, accurately repeated in *2 Henry IV*. Potter, a remarkably precise reader and writer, claims that scenes in Part 2 "*largely* repeat" (emphasis mine) scenes from Part 1. The qualifier "largely" is necessary because the repetition ranges from imprecise to impressionistic. To begin to understand how *2 Henry IV* inaccurately repeats *1 Henry IV* is to begin to understand how the play works on its own terms as a work of art obsessed with what Lady Percy called "recordation."[39] Throughout part 2, lines and facts from part 1 and from *Richard II* tend to be misremembered or misrepresented. In act 3 of *2 Henry IV*, for instance, King Henry IV recalls the occasion when the deposed Richard II admonished Northumberland (although Northumberland and his son, Hotspur, eventually rebelled against Henry IV, their support played a crucial role in deposing Richard II). He even recalls the prophetic warning that Richard spoke against Northumberland. Or, rather, he recalls something that he seems to think was Richard's prophetic warning against Northumberland. Looking to Warwick, Surrey, and John Blunt, the king asks:

> Which of you was by—
> [to Warwick] You, cousin Nevil, as I may remember—
> When Richard, with his eye brimful of tears,
> Then checked and rated by Northumberland,
> Did speak these words, now proved a prophecy?
> "Northumberland, thou ladder by the which
> My cousin Bolingbroke ascends my throne"—
> . . . .
> "The time shall come," thus did he follow it,
> "The time will come that foul sin, gathering head,

Shall break into corruption"
(3.1.65–72, 76–78)[40]

Henry IV's memorial reconstruction is a surprisingly accurate quotation from *Richard II*. Compare the following (direct correlations are highlighted in bold):

Northumberland, thou ladder by the which
My cousin **Bolingbroke ascends my throne**—
(*2 Henry IV*, 3.1.71–72)

Northumberland, thou ladder wherewithal
The mounting **Bolingbroke ascends my throne**
(*Richard II*, 5.1.56–57)

More shocking than the fact that Henry IV so accurately remembers Richard II's words is the fact that he never heard them in the first place. Henry IV was not present when those lines were spoken by Richard, nor were any of the characters he asks to confirm his memory (Warwick, Surrey, and John Blunt).[41] Richard spoke those lines in the presence of only three other characters: Queen Isabel, Northumberland, and a guard. The lines that Henry IV was actually witness to, which is to say the lines that he should have recalled, were these from the beginning of act 4:

Gentle Northumberland,
If thy offenses were upon record,
Would it not shame thee in so fair a troop
To read a lecture of them? If thou wouldst,
There shouldst thou find one heinous article
Containing the deposing of a king
And cracking the strong warrant of an oath,
Marked with a blot, damned in the book of heaven.—
Nay, all of you that stand and look upon me
Whilst that my wretchedness doth bait myself,
Though some of you, with Pilate, wash your hands,
Showing an outward pity, yet you Pilates
Have here delivered me to my sour cross,
And water cannot wash away your sin.
(*Richard II*, 4.1.239–53)

These lines that Henry IV heard with his own ears are memorable lines with biblical cadences (and metaphors), especially "Marked with a blot, damned in the book of heaven" and "And water cannot wash away your sin." Yet instead of these lines, King Henry quotes words he never heard. Is Shakespeare or is Henry IV misremembering Richard II? James Bulman suggests that this confused recollection "contributes to the network of reports by which this play interrogates the nature of historical truth and invention."[42] But Henry IV's impossibly confused recollection also contributes to a network of misremembering, of "delible" or "delete-able" records.

Act 4 of *2 Henry IV* moves the drama to Gaultree Forest, an Anglicized place-name that recalls "one of the principal traditional constituents of ink, derived from oak-galls, which are excrescences produced on oak trees by the action of insects."[43] There, in the midst of what should be a crucial battlefield parley, Westmoreland and Mowbray suspend negotiations to argue about the tournament between Bolingbroke (now Henry IV) and Mowbray's father, one of the crucial plot points that set into motion the action of *Richard II* and *1 Henry IV*. Mowbray maintains that his father would have killed Bolingbroke/Henry IV if the tournament had not been stopped by Richard II. "You speak, Lord Mowbray, now you know not what" (4.1.136), replies Westmoreland before arguing that if Bolingbroke had not killed the elder Mowbray, then the people, who "doted on" Bolingbroke and "cried hate upon" Mowbray, would have done the job themselves (4.1.145–46). The conversation is, as Westmoreland realizes, a "digression from my purpose" (4.1.149)—an imaginary rewriting of history along the lines of "what if" at a moment when both men's lives are in danger. Eventually, the Archbishop, who leads the rebels, convinces the others that a peaceful surrender is a viable option, for, as he supposes, the king will be glad to end the matter by forgiving and forgetting. According to the Archbishop,

> the King is weary
> Of dainty and such picking grievances,
> For he hath found to end one doubt by death
> Revives two greater in the heirs of life;
> And therefore will he wipe his tables clean
> And keep no telltale to his memory
> That may repeat and history his loss
> To new remembrance.
> (*2 Henry IV*, 4.1.207–14)

The passage is one of two references to erasable tables in *2 Henry IV*—the only two such references in the Henriad. In their essay on the material and metaphorical function of writing tables in Renaissance England and especially in *Hamlet,* a play that, like *2 Henry IV,* "obsessively stages erasures of memory," Peter Stallybrass, Roger Chartier, John Franklin Mowery, and Heather Wolfe demonstrate that although writing tables are "a technology of memory, the tables are also a technology of erasure."[44] It is fitting that such a revisable device would be cited here, just before Prince John employs a dirty trick: breaking the spirit if not the letter of his promise to dispatch of the insurgents, he condemns the rebels to death. The news of Prince John's success, delivered to the king by Westmoreland, seems to please him at first, but then he weakens: "And wherefore should these good news make me sick? / Will Fortune never come with both hands full, / But write her fair words still in foulest letters?" (4.3.110–13). Henry IV is carried to a bed, from which he never rises.

King Henry IV's death, too, involves narrative revision. Disapproving as he was of his son's transgressions, his own record was marked with a blot, "the deposing of a king," to quote Richard II, the king he deposed (*Richard II,* 4.1.244). To atone, he had vowed at the end of *Richard II,* "I'll make a voyage to the Holy Land / To wash this blood off from my guilty hand" (5.6.49–50). *1 Henry IV* begins with a similar vow, slightly altered: the king says nothing of washing guilt away, but rather of a duty to spill the blood of pagans in the Holy Land. "Our purpose now is twelve month old," says the king, and in that time, it seems his purpose has changed (*1 Henry IV,* 1.1.28). More political unrest ensues, and he dies in *2 Henry IV* before making a penitent journey or crusade to the Holy Land. But he dies in a room named Jerusalem, and so he makes one more slight revision to the story: it was less a plan by him, more a prophecy about him: "It hath been prophesied to me many years, / I should not die but in Jerusalem, / Which vainly I supposed the Holy Land" (*2 Henry IV,* 4.3.397–99). In Folio 47, we have reached the page opening with which this chapter began, and the blot is hard to ignore at the bottom of the page. There it spreads, reminding us that despite what politicians say they said or meant, present revision does not necessarily remove or cover up past records. Reading the blot in conversation with *2 Henry IV*'s mixed metaphors for inscriptive revision, we begin to understand that the technical differences between razing and blotting allow a better reading of Shakespeare's Henriad (and of numerous other texts in the period that take up the topic of writing and its relationship to remembering and forgetting).[45] And not just the differences, for the similarities between razing and blotting also allow for shades of contrast between two substrates that existed in apposition to each other in medieval and Renaissance England. It is to these differences and similarities, and to their material and metaphorical influence on the recording and revising of history in Renaissance England, that I now turn.

Razing Animals

Erasing was a routine task for Renaissance writers using animal-based substrates, but, as I discuss below, paper altered the revision routine in ways that have material and metaphorical significance. I begin with parchment, and I use the term "parchment" here as a catchall for any animal skin prepared as a writing surface, including vellum. Peter Beal claims that "the terms 'parchment' and 'vellum' have commonly been used interchangeably since at least the early sixteenth century," and *Hamlet* supports this claim:

> HAMLET. Is not parchment made of sheepskins?
> HORATIO. Ay my lord, and of calves' skins too.
> (*Hamlet*, 5.1.116–17)[46]

Technically, "vellum" is the skin of a calf, but as Hamlet and Horatio indicate, "parchment" was commonly understood to denote calfskins, too.[47] Like Hamlet, Jack Cade in *2 Henry VI* is aware of parchment's sheep origins (indeed, he is aware of the lamb's unwilling participation in an inscriptive economy that does violence to less legally protected bodies):

> DICK. The first thing we do, let's kill all the lawyers.
> CADE. Nay, that I mean to do. Is not this a lamentable thing, that of the skin of an innocent lamb should be made parchment? That parchment, being scribbled o'er, should undo a man?
> (*2 Henry VI*, 4.2.75–79)

Notably, the plan to "kill all the lawyers" is not simply a coup against a documentary legal system; it is also a re-appropriation—even a retro-appropriation—of tools that have been misused by the participants in that documentary culture. Dick the Butcher is the counterpoint to the clerk of Chartham who appears in the play just after the lines cited above (see 4.2.83). Knives were used not only in butchering an animal, flaying it, and making a writing surface out of its skin, but also in the act of inscribing and reinscribing the multilayered, membranous surface. Medieval images depict scribes using knives in their left hands to hold down and/or smooth out any cockles in the parchment on which they are writing.[48] In addition to serving as a tool to flatten the page and to cut and maintain a writer's quill pen, the knife also served as an eraser. Some scribes preferred two blades: one for pen maintenance and another for erasing. As a result, one might have a separate "erasing knife" or the penknife might have a built-in "erasing blade."[49] Understanding the scribe's knife as an eraser, anyone familiar with the texture of early paper immediately begins to understand how fundamentally different paper and parchment are as writing—and especially as revising—substrates.[50] Unlike paper, which has interlocked and overlapping fibers, parchment has layers. To put it another way, parchment has two surfaces plus a series of potential surfaces.[51]

The palimpsestic qualities of texts that fascinate modern scholars, equipped with infrared technology that makes under-texts legible, are qualities that could cause consternation for historical writers and readers. The primary concern was manipulation. In *Of Reformation* (1641), John Milton writes, "Who knows not how many surreptitious works are ingraff'd into the legitimate writings of the Fathers, and of those Books that passe for authentick who knows what hath bin tamper'd withall, what hath bin raz'd out, what hath bin inserted[?]"[52] One way to avoid unauthorized changes was to use sheepskin, which tended to yield a lower-grade parchment. Instructing modern scribes interested in historical practices, Michelle Brown and Patricia Lovett recommend against using it because it is greasier than vellum. As a result, the parchmenter has a harder time preparing it, and, significantly, the scribe has a harder time clearly removing mistakes from it.[53] This imperfect erasability—a drawback in most instances—was a virtue that made sheepskin ideal for use as legal indentures. According to *The Dialogue Concerning the Exchequer* (ca. 1180), "The writing of the roll has this in common with charters and other letters patent, that it may not be erased: and for this reason care has been taken to make them of sheepskins; for they do not easily yield to erasure without the blemish becoming apparent."[54]

Another way to avoid and catch attempts at unauthorized erasure and insertion was to avoid manipulable writing materials, such as carbon inks, which, according to L. C. Hector, "were not used for documents intended to have any legal validity." Given the wide variety of historical ink recipes available, many of which contain variant ingredients unfamiliar to modern readers, I find Hector's taxonomy especially useful: "Black inks of all periods can be divided into two classes, those which contain iron salts and those which do not." Hector further explains that standard medieval ink recipes call for galls and copperas or vitriol. The copperas or vitriol, both names for ferrous sulfate, are the iron salts. These iron salts react with the gallic or tannic acid in the galls, "produc[ing] a purplish-black compound which in normal conditions grows even darker with age." Ink (*incausta* in Latin) gets its name from this chemical reaction and from the resulting fixity of such ink when applied to parchment. As Hector writes, "The way in which they [inks] integrated themselves with the surface to which they were applied was conceived as a figurative 'burning in.'" Carbon ink, ink made from lampblack (soot) and mixed with gum water, was also used by scribes writing on both papyrus and parchment because it "provided an immediate and dense black which commended itself especially to the writers of ornamental headings."[55] Erasure of carbon ink is simple, and therefore problematic: "Carbon-and-gum ink remains largely upon the surface of the writing material. . . . While this allowed for easy erasure . . . abrasion could take an unwanted toll."[56] Despite its erasability, carbon supplied a richness of color that justified its use, and it was sometimes added to iron inks to enhance the color.

In his study "Erasures in Thirteenth-Century Music," John Haines embraces and perfectly illustrates the way modern scholars seek for that which historical scribes meant to hide. Describing his examination of the *Jena* manuscript, "the most important source for early Minnesang" (Middle High German love songs), Haines writes: "Thankfully for us, *Jena*'s note heads, at 2.5 x 3.5 mm, are larger than in most books, so that even when erased they are hard to miss. In general, however, scribes left us tell-tale marks, some quite generous. . . . Some trace of the act of erasing usually survives."[57] Here we, with Haines and perhaps historical readers from the thirteenth century onward, are the scribes' audience, the readers for whom they "thankfully" leave "tell-tale marks, some quite generous." From a scribe's point of view, however, the goal is to conceal errors and corrections. A successful scribe was an artisan whose tactile skill with the plant, animal, and mineral materials used to make texts allowed them to respond to the unique situation of an error. They knew that it was easier to hide a mistake on the hair side of a parchment page.[58] For instance, when Bernard J. Muir began editing *The Exeter Book* (a collection of Old English poems from the tenth century), he was surprised to find "over four hundred scribal alterations in the manuscript" not mentioned by previous editors "because they *are not visible* in the facsimile."[59] Many of these alterations involved "scrap[ing] away one or more letters and enter[ing] the new reading," and Muir claims that "most examples" of that type of alteration "have never been documented." These oversights, however, are not simply due to editors working with facsimiles. Even working with the original, Muir had to hold up the manuscript "at different angles to a light source" to find the scribes' careful corrections. Some of the corrections Muir found are invisible to the naked eye: they "must be detected by passing the hand lightly over the surface of the text and *feeling* for the slight abrasiveness left by the scraping."[60]

Blotting Plants

Even though the finest paper might have been preferable to the worst parchment—"scarred or blotchy" parchment from a parti-colored animal or parchment that is "cut, split, holed, torn, stretched, strained, or dried to a horn-like consistency"—the two substrates require fundamentally different approaches to correcting error.[61] Haines claims that "in the Middle Ages, to write . . . also meant to erase";[62] as paper became common in Europe, to write also meant to blot. Of course, some of the less aggressive ways of marking error transferred from parchment to paper. One might strike through on both substrates or use expunction: placing a series of dots under the words to be corrected. One might even use a penknife to score lines through an incorrect word. With ink that has not yet integrated with its surface, one might wipe up or soak out an error. But paper does not have the same layering that parchment does, so once ink had penetrated the surface and integrated with the paper's fibers, blotting was the best one could do to obliterate an error.

The earliest paper document in the Public Record Office is from ca. 1220, during the reign of Henry III, but paper records were not common in England until nearly three centuries later.[63] Hector claims that "by the end of the [fourteenth] century paper is almost as likely as parchment to be the material chosen for household accounts," a statement that recalls the end of act 2, scene 4 of *1 Henry IV*, when Hal and Peto search Falstaff's pockets and find "nothing but papers" on which Falstaff's expenses are itemized (2.4.553).[64] Manuscript copies of Chaucer show similar evidence of media mixing: "Of the eighty-three manuscripts of Chaucer's *Canterbury Tales*, all written in the fifteenth century, 61 percent are on parchment, 34 percent are on paper, and 5 percent are mixed."[65] According to Ann Blair, "The increased availability of paper coincided with a rise in paper records of all kinds, from diplomatic correspondence to family papers to collections of scholarly notes."[66] Indeed, according to Blair, "the principal demand for paper before the invention of printing stemmed from the production of legal and administrative documents, rather than the needs of readers or authors. Paper was cheaper than parchment, by a factor of eight at its first introduction in England, and by a factor of 32 in 1500."[67] It would be difficult to overstate the significance of Blair's observation in the context of a conversation about mixed media, documentary culture, and inscriptive error: the early adopters of a substrate that made error more rather than less visible were those charged with keeping official records.

"Paper was in use alongside parchment from the fourteenth century onwards and gradually became more and more common," claims D. C. Coleman, but despite this general mixing of media, parchment remained the preferred substrate for official records until the turn of the sixteenth century: "Not until Tudor times is paper frequently to be found amongst the records [in the National Archives]."[68] Coleman cites town accounts, records of guilds and colleges, personal diaries, letters, and merchant accounts as examples of the kinds of record keeping that increasingly relied upon paper, from the fourteenth century on. However, Coleman notes, "the old survived with the new: though the minutiae of correspondence and negotiations which were a part of the great boom in property dealings during Tudor and Stuart times were conducted on paper, the formal evidence in the shape of deeds of sale or mortgage was engrossed on large and unmanageable slabs of enduring parchment."[69] The apposition of paper and parchment substrates allows characters in Shakespeare to oppose them, imagining their own errors as erasable marks and others' as indelible blots. But the side-by-side existence of parchment and paper, the apposition of these plant and animal media materials, make the opposition of one with another possible in the first place. (The apposition and opposition of ebooks and "real" books are a useful modern analogy.)

There are subtleties, details, and writing practices worth considering that cannot be treated fully here. Writing tables, which could be wiped clean, were widely available. Schoolboys were steeped in the literature of ancient cultures in which writing and erasing on wax tablets and papyrus were well known. Graphite became common in the first half of the seventeenth century.[70] A mistake on paper, caught quickly enough, before the ink had burned through the layer of gelatin sizing and into the paper's fibers, might be wiped up or soaked into sand or blotting paper (unsized paper that allowed blots to be transferred). A word on paper might be scratched out or covered with a cancel slip. But the apposition of two kinds of recording surfaces, one on which an error might be erased and made invisible, the other on which an error must be blotted, is fundamental. Indeed, other textual technologies like washable tables and erasable graphite can be better understood within the context of this apposition. As Alan Stewart claims, "The raw materials of writing possessed vivid associations for their early modern users, in part no doubt because they often prepared them themselves."[71] The intriguing, permutating relationships between the material ecologies of writing and reading and their varying metaphorical associations leave much left to be explored and understood. To draw this chapter's exploration to a close, however, I turn back to blots on paper and specifically to the permutation of a key metaphor in what was arguably the most well-known poem in Renaissance England: Psalm 51.

"Blot out my transgressions"

The word "blot," like the medium paper, became common in the years between Oldcastle and Falstaff, between the reign of Henry IV and *Henry IV* parts 1 and

2. It "appears first in 14th cent.: no corresponding form is known outside English, and the word may be really connected with *plot n.*, or may unite a notion of *spot* with some words in *bl-*." The *OED* notes that *blot* has been likened to similar words in Old Norse (*blettr*), Danish (*plet*), and German (*bletz*), "but no normal phonetic relation to these words can be affirmed." It first appears in Chaucer—who was writing during the reign of Richard II—both as a verb ("biblotte," 1374) and as a noun ("blotte," 1386).[72] It is intriguing, but not surprising, that a new word appeared in the English language, a word that sounds like *plot* and *spot* joined together, about the same time a new writing substrate was becoming common in England. Almost immediately, the word indicates both a visible inscriptive error on a page and a glaring moral error in a person. When Shakespeare was writing the English history plays, the word "blot" barely appeared in the Bible, but by the end of his writing career, the Authorized Version or King James Bible (hereafter KJB) had made the word common to readers of the Bible.

The KJB translators used some form of the word "blot" five times more often than any previous English Bible—seven times more often if the Apocrypha is included.[73] The word "blot" (or a variant of it) never appears in the Coverdale Bible or in the Wycliffe Bible, the Bible produced during the reign of King Henry IV. "Blot" appears just four times in the Geneva Bible and only three times in the Bishops Bible. But in the KJB, some form of the word "blot" appears twenty-one times in the Old and New Testaments and another eight times in the Apocrypha. In the late sixteenth-century Bibles, like the Geneva and Bishops, where the word "blot" is used, it clearly indicates a spot, stain, or blemish.[74] By contrast, in many cases in the KJB the word "blot" denotes the opposite of the word or words that had been used in previous English translations; where earlier versions describe sin that is "wiped out," "razed," "erased," or "washed," in the KJB sin is canceled out with a blotchy stain, an unsightly reminder of the moral error.

For example, the first use of "blot" in the KJB is in Exodus 32:32, where Moses asks God to "blot me, I pray thee, out of thy Booke, which thou hast written."[75] It is a verse that echoes in the Henriad, particularly in *Richard II*, where Mowbray proclaims, "If ever I were traitor, / My name be blotted from the book of life" (1.3.205–6) and where King Richard II tells Northumberland he is "marked with a blot, damned in the book of heaven" (4.1.246). The Hebrew word in Exodus 32:32, where the KJB translators give "blot out," is *mâchâh* (transliterated). James Strong's *Concordance* defines the word as "prop. to *stroke* or *rub*; by impl. to *erase*."[76] Comparison to other translations in circulation when the KJB was created shows that the Hebrew word presented no

Table 1. Blots and Books in the Bible (Exodus 32:32)

Version	*Translation*
KJB	"blot me . . . out of thy Booke"
Douai-Rheims	"strike me out of the book"
Geneva	"rase me out of thy booke"
Bishops	"wype me . . . out of thy booke"
Wycliffe	"do awey me fro thi book"

conundrum to translators (see Table 1). Although the various translations use different words, each of the words used, except for "blot," is faithful to the sense of the Hebrew word *mâchâh* and the erasure or removal that it implies.[77]

More striking than the KJB translators' extensive substitution of "blot" in places where other translators had used words that indicated wiping out or erasure is the fact that the substitution was not total. In twenty of the thirty-six occurrences of *mâchâh* (or some form of it), the KJB translates the word as something other than "blot," even using the word "wipe" in some instances.[78] For example, 2 Kings 21:13 describes the punishment that God will bring on "the house of Ahab": "I will wipe Ierusalem as a man wipeth a dish, wiping it and turning it vpside downe." In such a context, blotting would not work with the metaphor of cleaning out a dish and turning it over so that it is entirely empty. Even an alternate meaning of "blotting" that implies dabbing or soaking up would change the prophecy. In this passage, God tells Jerusalem that they will be removed from the land, a prophecy whose fulfillment begins just three chapters later in 2 Kings 24:14 when Nebuchadnezzar besieged and "caried away all Ierusalem" into captivity in Babylon. In the case of 2 Kings 21:13, then, the action of wiping (not blotting) works best with the metaphor of cleaning a dish and with the Jewish history known to the translators. Clearly, as this example demonstrates, the KJB translators understood the Hebrew word and how it could be translated. And yet, over and over, they used a form of the word "blot" where the Hebrew scriptures and other English translations previously relied on the language of wiping or erasing.

It is worth observing that the KJB is hardly alone in updating the language of the Bible to align with modern writing technologies. If the Virgin Mary had been able to read, she would have read a scroll, and yet she almost always has a codex in Renaissance depictions of the Annunciation. If a modern artist were to paint the Annunciation in traditional fashion, Mary would be reading from a smartphone, and museumgoers who have been looking at Mary reading new technology for as long as they have seen images of the Annunciation would notice and talk about (and complain about) the artist's modernization. Similarly, in Exodus 32:32, the passage displayed on in Table 1, only the KJB uses "blotted out," but every single translation uses some spelling of "book." When Moses tells God to raze or wipe him out, the text from which he imagines being erased is certainly not a codex book, but a scroll.[79] So what we see in most of the other instances is a Bible with partially updated writing technology metaphors: scrolls have become books, but they are books that are erased, not blotted. Which is to say, they are parchment books. By the end of the sixteenth century, however, the Geneva Bible's "rase me out of thy booke" was misaligned with media technology.

In many instances where KJB replaces erasure with blotting, the translators appear simply to be modernizing the metaphors. But not always. I close by considering a highly influential blot in the Bible, one that links Shakespeare's bildungsroman-styled history of King Henry V with one of the earliest recorded histories of another famous king's coming of age, reign, and death: King David's. We can guess at the KJB translators' reasons for using "blot" in some instances where the act of erasure is suggested in the original. In these cases, broadly speaking, a writing metaphor is updated to reflect modern practices of correcting inscriptive error. In places where the metaphor is clearly indicating an action that is not related to writing, KJB translates the word *mâchâh* in a way that works with the metaphor. The dish in 2 Kings 21:13, is being wiped out, not blotted. Or, in Genesis 6 and 7, God uses a global flood to "destroy" (not "blot") "euery liuing substance . . . from off the face of the earth."[80] Proverbs 30:20 warns about the adulterous woman who "eateth, and wipeth"—not blotteth—"her mouth, and saith, I haue done no wickednesse." And yet the most well-known instances of "blot" in Renaissance England, the Penitential Psalms, depart from this logic, replacing washing metaphors with textual-blotting metaphors.

Even if early modern readers did not dutifully follow the Book of Common Prayer's prescribed readings, which would have put Psalm 51 before their eyes at least once every month, they could hardly have escaped familiarity with the passage. I quote Hannibal Hamlin at length to give some sense of its importance:

> The Penitential Psalms [Psalms 6, 32, 38, 51, 102, 130, and 143], and Psalm 51 in particular, played a vital role in the liturgy of the English Church as well as in the private prayers and devotions of individual Christians. Psalm 51 also had an impact on literature, especially the seventeenth-century religious poetry of John Donne, George Herbert, and others. . . .

> The language of this psalm, its singular metaphors and vocabulary, proved irresistible for Christian writers, who, during these decades of religious turmoil, struggled with the problematic nature of sin and repentance. Psalm 51 was also a text in which sixteenth- and seventeenth-century reformers, Protestant and Catholic alike, found the seeds of ideas that ultimately affected what they believed and how they worshiped.[81]

Indeed, the first verse, where David asks God, in the Geneva Bible, to "put awaie" his sin, was especially important and, again according to Hamlin, was "one of the most widely known [verses] in sixteenth-century England because of its peculiar involvement in the legal loophole known as 'benefit of clergy.' "[82] In brief, this loophole allowed a criminal to claim benefit of the clergy—and therefore less severe punishment—if they could recite the first verse of the psalm in Latin.[83] As a result, Psalm 51:1 came to be known as "the neck-verse" for its usefulness in saving one's neck (in the most famous instance, Ben Jonson's neck) from the gallows.[84]

In the Geneva Bible, the version that Shakespeare seems to have relied upon most heavily, Psalm 51 is prefaced by this explanation: "A psalme of Dauid, when the Prophet Nathán came vnto him, after he had gone in to Bathshéba."[85] The incident alluded to is the entrapping story Nathan told David (about a rich man who steals and eats a poor family's only lamb) in order to rebuke him for his actions with Bathsheba—a scene alluded to in *Henry V*, when King Henry sentences three traitors to the very death penalty they had advised him to dole out to another man. In Psalm 51, in the 1560 Geneva Bible, David prays: "Haue mercie vpon me, ô God, according to thy louing kindenes: according to the multitude of thy compassions put awaie mine iniquities. Wash me throughly from mine iniquitie, and clense me from my sinne" (Psalms 51:1–2). A printed marginal note linked to "Wash me" in verse 2 explains: "My sinnes sticke so fast in me, that I haue nede of some singular kinde of washing." There is little question what sort of revision is being called for here, especially after verse 7: "wash me, & I shalbe whiter then snowe." Geneva's "put awaie," "wash me," and "clense me" in 51:1–2 echo the language of the 1535 Coverdale Bible, 1537 Matthews Bible, and the 1539 Great Bible (all of which use the variously spelled but identical language of "do away," "wash me," and "cleanse me") and are echoed in the 1568 Bishops Bible ("wype out," "washe me," and clense me"). Similarly, Geneva's use of "put awaie" in 51:9 ("Hide thy face from my sinnes, and put awaie all mine iniquities") appears as "put out" in the Coverdale Bible and "wype out" in the Bishops Bible. In all sixteenth-century versions of Psalm 51, then, a reader finds King David asking God to metaphorically wash away his sin. The metaphor is consistent throughout the psalm: never does King David make reference to writing or to records. By the KJB translators' logic elsewhere, the word "blot" should not appear in this passage.

Even before 1611, David's sin was hypervisible in Renaissance England: his was a narrative of folly and repentance that was not erased but repeatedly rehearsed.[86] Relying on the language of paper revision, the KJB made David's sin even more visible by replacing the Geneva's "put awaie" in the first and ninth verses of Psalm 51 with "blot out." In the KJB, then, as in the Henriad, blotting and washing coexist. In Psalm 51:1–2, this apposition results in a contradictory metaphor: "blot out my transgressions" and "[w]ash mee throughly from mine iniquitie." Are these sins supposed to be blotted or washed? Later, David begs, "Hide thy face from my sinnes; and blot out all mine iniquities" (51.9). But the blotted history, like the lines covered by a blot on the page, is the most eye-catching and the most interesting. Recorded, transmitted, and translated over time, King David's history remained the subject of poetry: "Peruse (who list) *Dan Davids* perfect deedes, / There shal he find the blot of *Berzabe*" writes George Gascoigne in a poem titled "The lover encouraged by former examples, determineth to make vertue of necessitie."[87] Like Henry V, David is a king whose errors are part of his narrative.

If the KJB's use of "blot" makes more sense with a narrative like King David's where the visibility of error is more interesting than a white-as-snow-washed narrative, it also shifts hamartiology, the theological doctrine of sin, in subtle but significant ways. If Hamlin is correct that "the notion that David's repentance was a model for all Christians hardly needed to be stressed."[88] and David's idea of what God does with or to sin becomes generally accepted, then it matters very much whether or not God's approach to forgiveness is more like blotting than erasing. Would it be too extreme to read our way forward from this crucial translation decision to something like *The Scarlet Letter*, published in 1850 but set in the mid-1600s, where sin can be forgiven, but where a religious community believes itself justified in forcing the sinner to be forever marked by that sin? The profusion of blots in the Bible that began with the KJB, a profusion that is so clearly linked to the materiality of recordation and revision on paper substrates, demonstrates in remarkable fashion how textual ecology informs and inflects important conversations across a range of studies, from modern bibliography to historical theology. Blots, and the roles that media play in preserving the past, remain pertinent to crucial media conversations now, too—especially to conversations about privacy, big data, and "the right to be forgotten and to erasure."[89]

One goal of this chapter has been to demonstrate how much we can learn about Renaissance textual materials and literary metaphors (and ecology and historiography and theology) by reading blots, but another goal is to suggest that the accidents of texts are, in and of themselves, worthy of our critical attention. The insights afforded by attention to textual ecology range widely in their seeming significance, but each allows the anthropocentric mind to lean out a little bit and look back on the history of books with new perception. To consider the real messiness of textual remembering and forgetting, of using a goose feather and diluted wasp larvae to scribble on top of gelatinized plants, is to practice a more complex and more playful mode of analysis that is at once more deeply material and more deeply metaphorical in its scope. Attending to textual ecology, we are, to return to Walker's resonant language, pulled into a promiscuous mode of aesthetic exploration in which we allow ourselves the time (and pleasure) of reading a large, stick-figured blot that interrupts Hal's attempt to erase the record of his past sins. And we will, in many cases, find that reading something we do not know how to value (a blot, a husk of flax, a squished spider) eventually leads back to the sort of traditional textual discoveries and insights that we have long valued.

The next chapter continues in this promiscuous, interrogative mode by turning to one of the most important and also one of the most invisible of textual ecologies: animal sizing, the gelatin solution used to coat early handmade paper. A properly sized page should allow writers and margin-scribbling readers to add neat, blot-free lines of ink to a paper surface. As we will see, the differences between what should happen on the surface of the page and what does happen on the surface of the page reveal how much we have yet to understand about book use and survival in both the past and the present.

CHAPTER 4

Sizing Matters: Annotating Animals in Renaissance England

> Tanning and papermaking were often located near each other, both dependent on fresh water and a stream to take away the filthy effluent.
>
> —Simon Barcham Green, "Sizing at Hayle Mill"

Scholarly discussion of animal sizing tends to be relegated to the domain of the paper specialist, and without good reason. Animals in the form of sizing haunt the paper leaves we turn in archival libraries, and they record vital information that should matter to us if we care to understand book production, use, and survival in the Renaissance and beyond. Without sizing, writes Jean Imberdis in 1693, "you could not one Letter make, / Nor any Mark upon the paper's Back."[1] Or upon a printed book's margins, for that matter. Animal sizing, Imberdis's description suggests, is an affordance that "set[s] limits on what it is *possible* to do with, around, or via the artefact"—in this case, paper.[2] It is also a feature that, added to paper, dramatically improves the paper's chances of long-term survival; it is therefore a feature that should be, but is not, accounted for in book survival estimates. Throughout this chapter, I use the terms animal sizing, gelatin sizing, paper sizing, and sizing somewhat interchangeably, as does the literature on sizing. I default, however, to animal sizing because doing so reminds us of the hybrid nature of Renaissance texts and of the ecopoetic negotiations involved in even the simplest acts of writing.

Animal sizing transformed the absorbent surface of handmade paper into a surface on which one could write with water-based inks, for without the aid of boiled-down animal parts, early handmade paper was not viable as a writing substrate. In an issue of *Hand Papermaking* devoted to the importance of sizing for artisans across a span of centuries, Timothy Barrett shares a succinct definition of sizing that he learned as a young papermaker: "Sizing is a material added to paper to change its resistance to the penetration of a liquid."[3] In its simplest state, paper "is hygroscopic, which means that it is sensitive to moisture and therefore dimensionally unstable."[4] Paper will soak up water-based writing ink unless it is coated in a substance, like sizing, that yields a "harder surface to minimize its absorbency" and to keep ink from "spread[ing] among the fibers."[5] By bonding with paper's plant fibers, animal sizing created a surface capable of mediating a writer's ideas in the form of clean lines of handwriting. Quality handmade writing paper, then and now, allows a writer to execute sharp, unfeathered ascenders, crossbars, and flourishes.

The previous chapter considered animal substrates alongside plant substrates in Renaissance England, attending to the unique ways in which one might correct handwriting mistakes on either parchment or paper and showing how the material, substrate-dependent lexicon of blots and erasures was carried over into metaphoric language about personal error, historiography, and even hamartiology. This chapter considers the presence of animals on and within plant-based paper, calling attention to the hybridity of early handmade paper, a hybridity that, although often overlooked, has shaped the history and ecology of texts and their uses. As in the previous chapter, the raw materials discussed in this chapter are affected by ecological availability and scarcity, they play a unique material role in the history of books, and they come to have metaphorical significance in Renaissance literature. Unlike the flax shives discussed in Chapter 2 and the blots discussed in Chapter 3, however, animal sizing tends to be less visible on the page. And yet, despite its relative invisibility on the pages of early texts and in our accounts of book history, sizing matters to our scholarship because it is a precondition of book annotation in Renaissance England and a significant indicator of book survival into the present day.

Studies of the history of reading have generally taken it for granted that readers could write in their books: a printed page either is or is not annotated with manuscript notes, but its "annotatibility" is not questioned. When William H. Sherman analyzed "patterns of use" at the Henry E. Huntington Library, he found that 1,531 of the 7,526 STC books, or 20.3 percent, "contain manuscript notes by early readers (not just signatures, underlining, and nonverbal symbols but more or less substantial *writing*)."[6] Sherman suggests, however, that "the practice [of marking books with manuscript notes] must have been much more widespread" than his findings indicate, in part because "the more heavily a book was used, the more vulnerable it was to decay."[7] And yet, recent and ongoing research about the role of sizing in book survival suggests that the opposite may be true. Some kinds of books, especially large books that were meant to be heavily used, were less vulnerable to decay because their pages were generously coated with sizing; this coating, which allowed readers to annotate the volume with water-based ink, tended to preserve the book's pages over time.[8] In other words, annotations in a printed book indicate the presence of sizing, and the presence of sizing is an indicator for book durability. The data suggest that the more annotatable a book was, the more likely it was to survive. Certainly some books were read to death and have disappeared, but perhaps for a reason that book historians interested in viable survival estimates have passed over: they lacked sufficient animal sizing to protect them from decay. Looking more closely at histories of papermaking and at early handmade paper itself, we begin to see that animals in the form of sizing played a key design role in the history of media, for their presence or absence defined how early handmade paper could be used, and, to a great extent, determined how long that paper would last.

Seeing Sizing

How, then, does one learn to see this unseen ingredient on early handmade paper? Even Renaissance writers and readers did not trust their eyes when it came to sizing, testing it instead with their ears and/or mouths. According to Imberdis's seventeenth-century poem on papermaking, "If you seek a surface fit for Pen, / First see if Size sufficient it has ta'en"; the sure tests of a well-sized page are these:

> Shaken by hand, it does not limply sag,
> As is the fashion of a piece of Rag,
> But crackles parchment-like, resisting still
> The holder's Touch, and loth to do his Will.
> If you seek further proof, then take the Sheet,
> And, licking it, with Spittle make it wet;
> If on the back no moisture-trace is found,
> Be sure, with ample Glue the Paper's bound;
> If it show through, the lack of Size is plain,
> On porous Paper you will write in vain.[9]

Modern readers, unwilling to risk the scandal of being dragged from a library reading room, may be forgiven for not licking archival documents, though it is worth noting that Imberdis's spittle test is a simple version of current sizing assessment practices that test for "surface wettability."[10] The hand-and-ear test is easier with loose leaves than with paper bound into a book, but anyone who has turned the pages of an expensive, well-made sixteenth-century folio can probably recall the crackling sound of turning a thick, crisp page. As Imberdis notes, this "parchment-like" crackle indicates a parchment-like substrate, one whose porous nature has been remedied by the application of animal sizing. Handwriting is the most visible indicator of sizing; historical papers bearing legible, clean lines made with water-based ink attest to the fact that the paper's porous surface has been coated in some way.

While handwriting is a reliable indicator of sizing, blotchy handwriting sometimes reveals a page that is poorly or unevenly sized. A good example of how uneven sizing affects annotatibility can be seen in the Newby Bible, a 1580 Geneva Bible at the Folger Shakespeare Library filled with handwritten marginalia.[11] According to Katherine Acheson, "In addition to the ownership marks from the seventeenth century [handwritten throughout the book], the Newby Bible has over 600 marginal notes," written by Elizabeth Boggis, that "record her experience of the world of Methodism in late eighteenth-century London."[12] In one marginal note dated November 27, 1791, next to Psalm 119, Boggis ran into a weak spot, a blemish in the paper where the sizing was inadequate. (As I discuss below, a number of environmental and logistical factors related to immersing small stacks of paper in a warm gelatin bath, separating those sheets by hand, and hanging them to dry could cause some areas of a sheet to absorb less sizing than others.) The annotation, written vertically along the outer margin of the page, explains that the regular minister was "at Bath on account of ill Health." A substitute preacher, Boggis writes, took "his text from Psalms 119 verse 75." However, the marginal annotation is not quite that clear. A diplomatic transcription would be "his text [. . . .] [. .]~~om~~ from Psalms 119 verse 75," but replacing the bracketed periods with some blot-like emoji would yield a more accurate representation, for it appears that when Boggis first attempted "from," her ink sank into a rough section of the paper. Boggis tried the word again, but half of it sunk, so she crossed out what was left ("om") and, back on a well-sized portion of paper, was finally able to complete a blotchy, but legible "from."[13]

To write about a word that has sunk or is sinking into the page is to recall the vocabulary used by Renaissance writers to describe their negotiations with the variously absorbent surfaces of heterogenous, handmade paper.[14] A 1645 pamphlet tirade against error-filled Bibles complains about the thin, spongy paper used for printing: "very course and extreme thin and bad *sinking-paper*."[15] The concern in this case is material, but sinking paper is used as a poetic trope throughout the period, too. A funeral sermon printed in 1626 promises that God will write our works in royal paper, therefore we should not "let . . . feare inuade vs, as if that paper could sinke, and so wee should lose our workes."[16] Here God, who is at other times invoked to blot sinners' names from the book

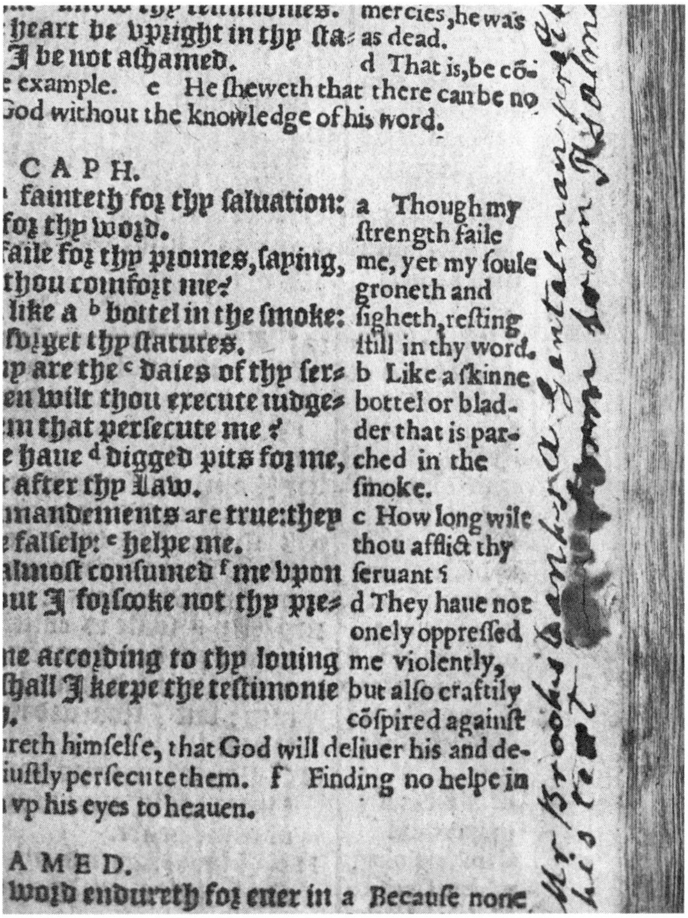

Figures 20 and 21.
Cover and detail of
sinking paper from
"The Newby Bible"
(1580) STC 2129,
sig. Ii3r. Used by
permission of the Folger
Shakespeare Library.

of life, is said to record good deeds so carefully that it is as if he is writing on high-quality, blot-resistant paper. Elsewhere, "sinking paper," a clergyman claims, is "as able to containe the distinct impression of the letters written thereon, as corrupt reason can distinguish or discerne the stampe or notions of holy things."[17] Put another way: a corrupt mind can see truth about as clearly as one can read letters written on sinking paper. It should come as no surprise that the language of blotting and of sinking paper is used in religious texts, for the Protestant Reformation thrived on readerly interaction with Bibles printed on paper. Animal sizing, a material precondition for such interaction, is what allowed—or disallowed—a reader/writer like Boggis to annotate her Bible. It is also, as I demonstrate in more detail later in this chapter, the material basis for a sinking paper trope that appears throughout sixteenth- and seventeenth-century texts, including religious treatises and printed sermons, scientific texts, schoolbooks, political pamphlets, poetry, and drama.

This chapter works to fill gaps in our knowledge by making visible a largely invisible feature of early handmade paper; I proceed in sections that consider the role of animal sizing first in paper production, then in paper use (materially as well as metaphorically), and finally in paper preservation. Failure to account for sizing in the stories we tell about the production and use of material texts, then, keeps us from better comprehending and describing historical reading practices and figurative language based on, especially, the sponginess of paper, and it bars us from accurately estimating the survival rates of Renaissance books. Easy as it is for us to overlook animal sizing, when we begin to look for it, we realize it mattered a great deal to those who made and used texts in Renaissance England. Seeing it, we can make more sense of human reading and annotating habits and we can better understand the nonhuman habitats that allow evidence of those human habits to survive to the present day in the form of paper.

Production, Phase One: Making Hybrid Media

Two key facts might be thought of as pillars that support the production history that immediately follows and that grounds this chapter's subsequent sections on use and survival: (1) pulp becomes paper, by definition, after it is pressed and dried and before it is sized, and therefore (2) adding sizing to paper was a separate process that required a series of separate, additional steps. Once we understand that the application of sizing is a nonrequired, additional step in the papermaking process, we begin to grasp the economic significance of sizing, and therefore the potential for a degree of paper heterogeneity that we have not fully accounted for in history of the book scholarship. We might think of gelatin sizing as an upgrade, a value-added or premium feature (a feature that contributed to the "enormous diversification" of the book market that Andrew Pettegree has convincingly demonstrated throughout *The Book in the Renaissance.*)[18] Article- and chapter-length discussions of animal sizing may be found in book conservation scholarship and in publications like *Hand Papermaking* that focus on the interests of papermaking artisans, but book history scholarship lacks a sustained consideration of animal sizing. One result is that the scholarship on sizing has tended to focus on practical questions of production and conservation. Drawing together the interests of literary criticism, book history, and the environmental humanities, I call special attention to how sizing has affected historical book use. I begin with a somewhat detailed description of the sizing process at the production stage, gathering what is known—and gesturing toward what is not known but that might be further explored—about the production of sized sheets of writing and printing paper.

In *Vulgaria Uiri Doctissimi Guil. Hormani Caesariburgensis*, William Horman's book of Latin lessons for schoolboys printed by Richard Pynson in 1519, we find descriptions of the various kinds of paper, including "cours papyr / that wolde serue for no wryttynge: but for marchandis / and pedlars to wrappe thyr stouffe in" and "Blottynge papyr [that] serueth to drye weete wryttynge."[19] The Latin text beneath the English refers to "Blottynge papyr" as "Charta bibula," bibulous paper, paper that imbibes moisture. (Unsized paper is often referred to now as "waterleaf," a word that was not in usage until the end of the eighteenth century.[20]) Such paper had been pulled from a vat of fibers, pressed in order to expel excess water, then hung to dry. The process is illustrated in detail in Joseph Jérôme de Lalande's *Art de faire le papier*, published in 1761 and acknowledged as the first technical account of the process of papermaking, and in Jean-Baptiste le Rond d'Alembert and Denis Diderot's *Encyclopédie*, published in 1767 and quite possibly based on Lalande's illustrations.[21] In Figure 22, the vatman at the far left pulls sheets from the vat to be stacked and then pressed; in Figure 23, the sheets are hung to dry. The dry sheets were, by every definition, paper; they were useful as such for various purposes like blotting, wrapping, and, as I will argue, for printing, but not for writing upon.

If the paper was meant for writing, then it needed "surface sizing" ("internal sizing," sizing added to the pulp before the paper is made, was not common until the 1830s[22]). The dried sheets of absorbent paper in the drying loft were gathered into small stacks or spurs and taken to the sizing room, where they were "tub sized": soaked in a warm gelatin solution rendered from animal bones and hides. Next, the paper was pressed a second time, this time to expel excess sizing, and then hung to dry a second time. These additional steps meant

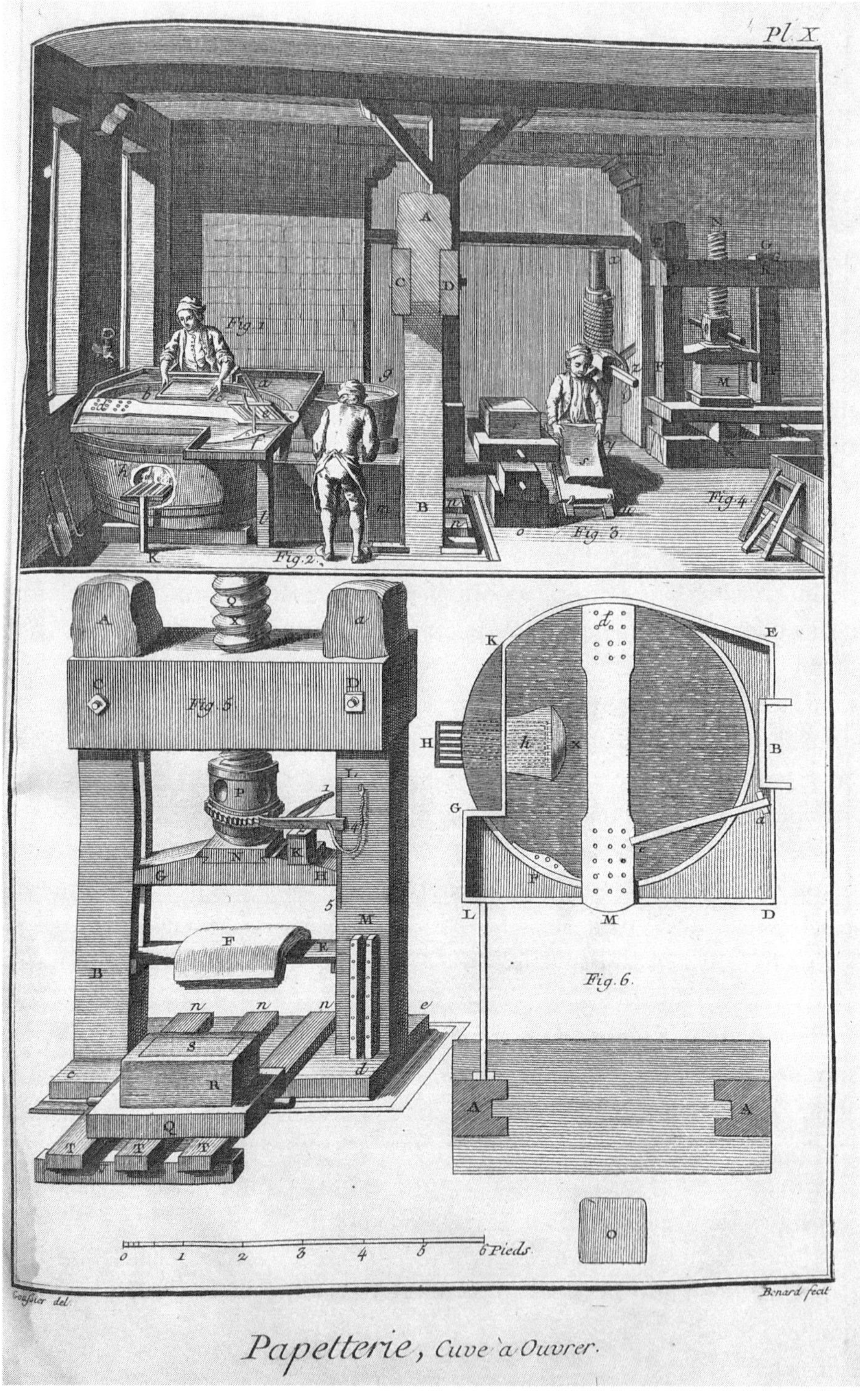

FIGURE 22. Making paper as depicted in Jean-Baptiste le Rond d'Alembert and Denis Diderot's *Encyclopédie*, vol. 5 (1767). Courtesy of the Department of Special Collections, Memorial Library, University of Wisconsin–Madison.

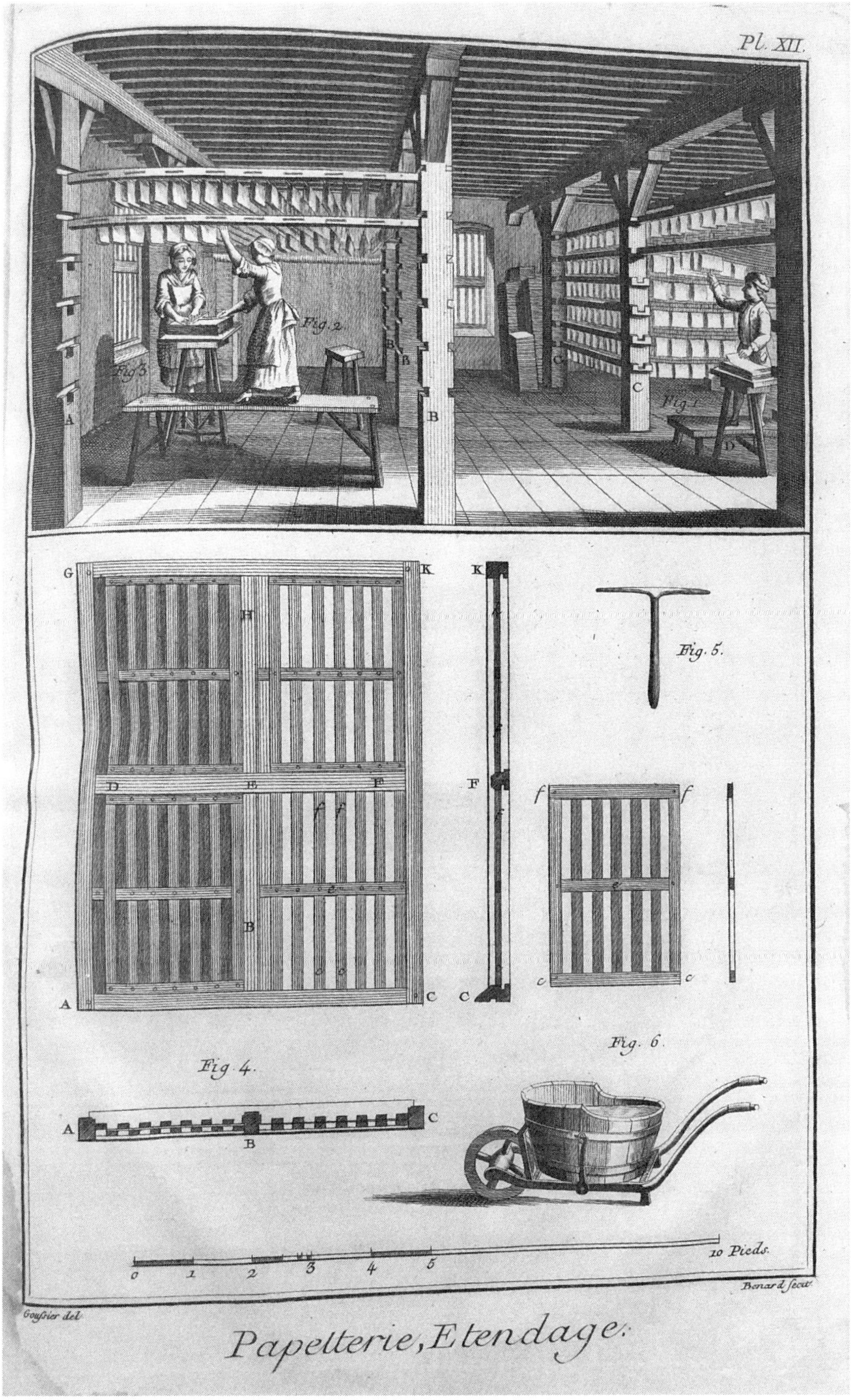

FIGURE 23. Drying paper as depicted in *Encyclopédie*, vol. 5 (1767). Courtesy of the Department of Special Collections, Memorial Library, University of Wisconsin–Madison.

additional effort and drying time, of course, but they also required additional equipment and space. In *Encyclopédie*, we see that the room for making paper and the room for sizing paper required similar equipment (vats and screw presses) and labor (three workers in each space).[23] A detailed inventory of a Southampton paper mill in 1696 argues for the accuracy of these illustrations: according to the inventory, the mill had a separate "Sizing Roome" with "2 Great Copper furnashes," "2 Great Sizing Tables," "1 Sizing press," and "1 washing Tubb," among other furniture.[24] Significantly, in addition to the added cost of gelatin sizing and the work that went into preparing, applying, and redrying the paper after it had been sized, there was also an added cost of loss. In the process of papermaking, if the vatman (the person using a mould and deckle to pull sheets from the vat) or the coucher (the person transferring the sheet from the deckle to the stack that will later be pressed and then hung to dry) were to damage a newly made sheet of paper, the damaged sheet—wet pulp at that stage—could be returned to the vat and re-formed into a new sheet.[25] If a sheet of paper were to be damaged during the sizing process, however, the sheet would be ruined. According to Dard Hunter, the sizing phase of papermaking "was extremely wasteful, as many sheets were torn and bruised beyond use."[26] He goes on to explain that the sizing room in early paper mills was known as the "slaughter house."[27] Such terminology is provocative: the slaughtered animals' body parts make sizing, which in turn "bruise[s]" and "slaughter[s]" the sheets of paper in the process of coating them.[28] Even sheets that made it through the tub-sizing process without damage might not be salable in the end, notes Cathleen Baker, because of "uneven sizing, sheets that stick (or 'block') together and cannot be separated, and mildewing or mold growth."[29]

Sizing pot ingredients were determined largely by resource availability, by the sorts of offal that papermakers could regularly depend upon.[30] *ABC for Book Collectors* defines size as "a thin glutinous or viscid decoction of bones or animal substances."[31] Philip Gaskell refers to sizing as "a solution of animal gelatine made from vellum or leather shavings boiled in water."[32] One also finds historical reference to parchment-scrap glue in John Florio's Italian-English dictionary (1598) under the entry "Carnizzo, *sinewes or threds of parchment to make size or glew.*"[33] However, Hunter defines early European sizing as a "gelatine rendered from the hoofs, hides, and horns of animals"—that is, from butcher scraps instead of tanner scraps.[34] Imberdis suggests another animal ingredient for the sizing pot—ears: "A store of Ears from Oxen and from Sheep / For just this very purpose men do keep"; Imberdis also claims that "some choose the Guts and leave the Ears aside."[35] Hunter credits Imberdis's volume as the "earliest European book to mention the actual sizing of paper,"[36] but he seems not to have known about an account of sizing that predates Imberdis's by two hundred

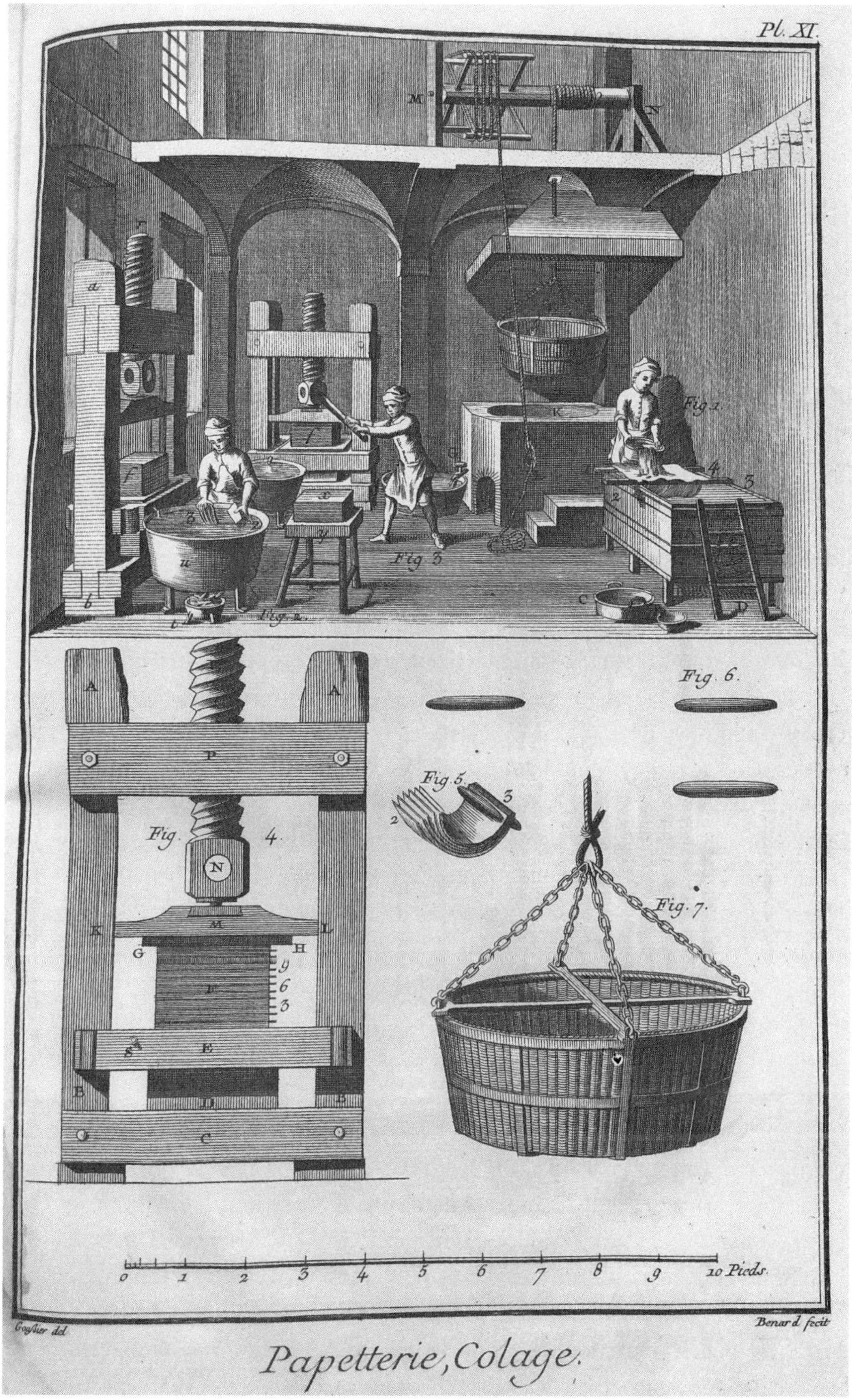

FIGURE 24. Sizing paper as depicted in *Encyclopédie*, vol. 5 (1767). Courtesy of the Department of Special Collections, Memorial Library, University of Wisconsin–Madison.

years: Francesco Maria Grapaldo's *De partibus aedium*, printed in Parma, Italy, in 1494. Grapaldo, like Florio, mentions leather and parchment parings as the raw material for sizing: "Later [the dried pages] are steeped in glue made from the waste scraps which tanners and parchmenters save for this purpose; they are hot-dipped, dried, and glazed, rendering them apt to take the pen and not to soak up the ink."[37] John Bidwell, writing about seventeenth-century English papermakers, claims that one of their challenges was "to learn how to brew a gelatin size solution, and how to apply it evenly, so that each sheet would absorb the right amount to bear ink properly and attain the desired printing and handling properties."[38]

The different solutions are alike in this regard: they make visible the ways that papermaking relied upon available natural resources, especially parchment/leather scraps and butcher scraps. The rag-and-bone collector is well known from history and literature; although most references to such a character focus on rag collection, bone collection was equally vital (and is preserved in the name).[39] The lack of standard sizing recipes may well indicate that papermakers knew they could throw into the sizing pot whatever animal parts were handy—hooves, hides, ears, horns, heads, parchment and leather scraps, bones. And yet it is clear that, over time, sizing practices grew more sophisticated and demanding. In addition to making paper that meets the basic requirements of writers, there would have been an added financial incentive to develop advanced techniques for those who could afford higher qualities of paper. Lalande reveals that, at least by the mid-eighteenth century, papermakers were perfecting the craft of sizing—or at least using the various animal scraps more discretely. The kind of animal used to render sizing was also a means of diversification and therefore a factor in determining paper quality and cost. John Bidwell notes that "writing grades [of paper] cost twice as much as printing grades because they required greater skill, better rags, a brighter colour, a smoother finish, and a harder surface."[40] One sizing ingredient might be valued for strength (tan-leather scraps), another for whiteness (sheepskin), and even isinglass, rendered from sturgeon, was used in rare instances.[41]

The most extensive (and recent) research on the complexities of animal sizing comes from Timothy Barrett, who, from 2007 to 2010, served as principal investigator for a team of researchers studying "how changes in gelatin and alum concentration influence the stability of historical papers."[42] This research, recently published as a beautiful, impressive project website titled Paper Through Time: Nondestructive Analysis of 14th- Through 19th-Century Papers, "analyzed 1,578 historical papers made between the fourteenth and the nineteenth centuries," employing nondestructive methods of analysis such as UV-Vis-NIR spectrometry ("for establishing the percentage of gelatin content and characterizing color") and XRF (X-ray florescence; "for evaluating the concentrations of residual metals").[43]

In the "Conclusions" section of the project site, Barrett writes, "All papers are best thought of as a system (both chemical and physical) impacted by a complex mix of internal and external variables."[44] His study takes many of these variables into consideration at an impressive level of specificity and detail. For instance, the "Discussion" page on the website carefully identifies potential "other components" that might cause paper discoloration, and in the "Plot Library" Barrett and his team visually represent the variations of gelatin content as well as (and in relation to) concentrations of iron, calcium carbonate, chlorine, potassium aluminum sulfate, and aluminum sulfate in historical papers.[45] Because *Paper Through Time* is freely available online, I do not reproduce any of its nearly one hundred plot graphs, but it is worth noting the sheer number of variables that Barrett and his team take into consideration (and this in a study that openly and repeatedly points to the limitations and artificial constraints of the research and especially of the data sample).

While the team was especially interested in chronological comparisons, the tools that they used allowed them to observe the effects of the sort of sheet-to-sheet variation that might happen even within the same paper mill. In a subset labeled "Same Book, Different Papers," the team explains that "in 229 of the books tested, we found two or more papers in clearly different condition, usually with one noticeably darker than the other. . . . Sometimes both sheets appeared to be from the same moulds, but their condition was clearly different." For an explanation, they cite Cathleen Baker's work on nineteenth-century hand and machine papermaking in America, which distinguishes "Monday paper" (dipped in a fresh pot of sizing) and "Saturday paper" (dipped in week-old sizing).[46] Part of the allure of Barrett's work is that it answers questions with facts and figures. And part of the allure is that, for all its answers, it raises more questions about the user experience of variously sized paper. An important insight about historical usage emerges from the charts, figures, and summaries: it is almost impossible to overestimate the heterogeneity of handmade paper produced in the first few centuries of European papermaking. Even the process of drying sized sheets could affect the outcome: "Great care was taken to ensure slow drying of the paper. Otherwise, it was believed, the size would not migrate to the surface of the paper with the moisture and the dried sheet would act as if only slightly, or not at all, sized."[47]

Production, Phase Two: Printing on Hybrid Media

While sizing rendered from varied body parts of varied animal species served writers' needs, it made more work not only for papermakers but also for printers. Like the papermaker, who had to add a series of steps to the process of papermaking in order to produce a sheet that could accept handwriting (dipping in gelatin, repressing, and redrying), the printer using sized paper had to add the extra step of moistening or misting the paper—of "melting" the sizing—in advance of printing so that the paper would accept oil-based printing ink. James Mosley calls attention to the fact that paper's fibers had to be softened by dampening "in order to make a satisfactory print on the hard-surfaced paper that was designed for use with the pen."[48] According to Lotte Hellinga, "Because oil-based ink was used, it did not penetrate the paper to any great extent; the paper had to be dampened to melt the size if the ink was to make any mark at all."[49] Sizing transformed a bibulous sheet of paper into a writing substrate that could accept ideas scrawled in water-based ink, but to be impressed with oil-based printing ink, the once-absorbent, now impervious sheet needed to be made more absorbent again. What is more, the use of dampened paper had the effect of dampening the wooden furniture of the printing press, and this affected how tightly the printer could lock up the chase—that is, tighten all the movable type to make them stay firmly in place—during printing. According to Ronald McKerrow, "Whether cracks in a block show in the prints from it seems to depend almost entirely on how tightly the block was locked up in the chase, and this, with wood furniture, must have depended a good deal on whether the furniture was dry or damp."[50]

The question naturally arises, if printing on sized paper was both more expensive, more time-consuming, and more burdensome on the printer's equipment, why not just print on unsized paper? Hunter claims to have observed incunables printed on unsized paper: "In an examination of more than a hundred different volumes of incunabula (books printed in Europe before 1501) I have found that . . . many of the sheets are heavily sized, others contain a limited amount of sizing, and a small number no sizing at all."[51] Describing the tools of the early printers, Theodore De Vinne backs up Hunter's claim, adding to it the assertion that not only could one print on unsized paper, but it was cheaper as well:

> The paper made for the *Bibles* of Gutenberg and for the earlier books was the ordinary writing paper of the period. . . . But the qualities which commended the paper to the copyist were objectionable to the printer. The hard surface caused harsh impression, and strong sizing made the damp sheets stick together. It was soon discovered that unsized paper, which, according to Madden, was about half the price of the sized, was easier to print. It would take a clearer impression, and more thoroughly imbibe the oily ink.[52]

Given the directness of these claims by two well-regarded experts, and given the impact that printing on unsized paper might have had on book use and book survival, it is surprising that these claims have not been taken up and either refuted or explained.

Some of the most convincing proof that oil-based printing ink could be effectively pressed directly onto unsized paper comes from the history of bookbinding. In *The Archaeology of Medieval Bookbinding*, J. A. Szirmai claims that "while sizing the paper was a prerequisite for preventing the writing ink from bleeding, it was not essential for printing and in fact many incunabula have been printed on unsized paper." According to Szirmai, printed incunables arrived at the binder on unsized paper; however, end-user expectations required binders to add sizing. Significantly, he describes binders adding sizing to the paper of already-printed books as a "remedy" for a commodity that was, in its unsized state, "unfit" for common reading practices: "Since unsized papers were unfit to write on and make glosses, a widespread practice in those days, it became the task of the binder to provide the remedy: sizing the printed sheets by soaking them in a warm solution of animal glue." The practice of sizing printed pages in the bindery "seems to have become routine in the last quarter of the fifteenth century; early manuals describe it in detail, together with the subsequent folding of the sheets."[53] If, as Bernard C. Middleton claims, binders began making and using glue from bones and hides in the early sixteenth century, it is plausible that they were making sizing as well.[54] Indeed, Szirmai suggests the direction of influence may be reversed, and that the production of animal sizing in binderies may have inspired innovative binding practices that relied upon glue.[55] In a study of how binders "coped with the printing press," Nicholas Pickwoad states, "For several decades following the introduction of printing, binders attempted to continue what had been traditional practices (such as elaborate sewing structures, wooden boards and metal furniture) until the pressure from the presses became too great and something had to give."[56] An increase in the number of binders might have managed the increased demand at first, according to Pickwoad, but ultimately, "the economics of the trade demanded something more fundamental."[57] One solution Pickwoad does not consider is the shifting of sizing duties from binders back onto papermakers. The economics make sense. Papermakers would have been responsible for the sizing of writing paper already, so the apparatuses for sizing printing paper were already in place.

A visual record of books printed on unsized paper in the sixteenth century appears in a well-known schoolbook, Charles Hoole's translation of Johannes Comenius's *Orbis Sensualium Pictus* (1659). The volume, a Latin-English grammar that offers, according to the title, "A *Picture* and *Nomenclature* of all the chief Things that are in the world; and of Mens Employments therein," has pages describing all sorts of objects and activities and includes a grouping that progresses from "Paper" to "Printing" to "The Book-sellars Shop" to "The Book-binder" and finally to "A Book." Numbers in each illustration are keyed to bilingual descriptions. "The Book-binder," for instance, depicts a bindery on the verso with Latin text and an English transcription on the recto (see Figure 25). The English transcription describes the scene: "In times past they glued Paper to Paper, and rolled them up together into one Rowl, 1. At this day the Book-binder bindeth Books, whilst he wipeth 2. over Papers steept in Gum-water; and then foldeth them together, 3. beateth with a hammer, 4. then stitcheth them up, 5. presseth them in a Presse, 6. which hath two Scrues 7 glueth them on the back, cutteth off the edges with a round Knife, 8. and at last covereth them with Parchment or Leather, 9. maketh them handsome, and setteth on Clasps. 10."[58] The text presents an idealized description here and elsewhere, a description meant to educate a student in Latin, not in bookbinding techniques. So, for instance, if step 9 (covering with parchment or leather) is optional, step 10 (adding clasps) is fully ornamental. The bindery in *Orbis Sensualium Pictus* depicts a conglomerate of space and time, of geographic and chronologic binding practices. It is presented as a modern description ("In times past . . . At this day") in which scroll and codex are contrasted. Bookbinding practices across centuries are merged in the depiction: the rounding and backing or "gluing up" that began in the sixteenth century (step 8); the setting on of ornamental clasps that would have required wood boards (step 10), not the pasteboard that was common in sixteenth-century bindings; and the sizing of printed pages (step 2), discussed in more detail below. Even a scroll, the textual object against which the binder's work is defined, is depicted in the foreground (labeled with a "1").

But it is step 2, and the depiction of this step, that I am most interested in examining. Hoole's translation states that "the Book-binder . . . wipeth over Papers steept in Gum-water." The illustration of this action, labeled "2" in the picture, shows four sheets of paper hanging on a line above the binder's head, near the ceiling. The image corresponds with an illustration of papermaking just three openings prior (see Figure 26).[59] There, papers are also hung at the highest point in the image. According to the description, "The Paper-maker . . . setteth them [the newly formed sheets] in the air that they may be dried."[60] The papers in the bindery are different: they are not blank. The pages, like those in the papermaker's shop, are hung to dry, having been sized (Hoole chose the word "gum-water" to translate "glutinosa," but the solution is animal, not vegetable).[61] The opening immediately after "The Book-binder" depicts "A Book." Or, rather, a number of books that have apparently completed the progression that *Orbis* depicts from paper to printing to bookseller to binder to bookshelf.

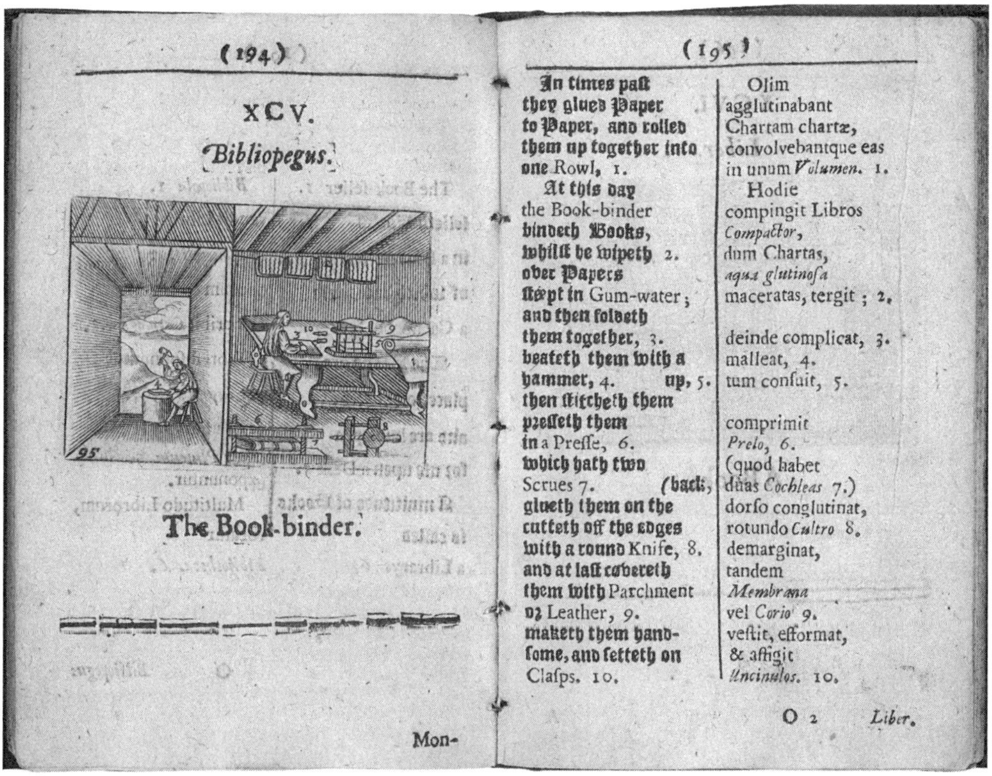

(194)

XCV.

Bibliopegus.

The Book-binder.

Mon-

(195)

In times past
they glued Paper
to Paper, and rolled
them up together into
one Rowl, 1.
At this day
the Book-binder
bindeth Books,
whilst he wipeth 2.
over Papers
steept in Gum-water;
and then foldeth
them together, 3.
beateth them with a
hammer, 4. up, 5.
then stitcheth them
presseth them
in a Presse, 6.
which hath two
Scrues 7. (back,
glueth them on the
cutteth off the edges
with a round Knife, 8.
and at last covereth
them with Parchment
or Leather, 9.
maketh them hand-
some, and setteth on
Clasps. 10.

Olim
agglutinabant
Chartam chartæ,
convolvebantque eas
in unum *Volumen*. 1.
Hodie
compingit Libros
Compactor,
dum Chartas,
aqua glutinosa
maceratas, tergit; 2.
deinde complicat, 3.
malleat, 4.
tum consuit, 5.
comprimit
Prelo, 6.
(quod habet
duas *Cochleas* 7.)
dorso conglutinat,
rotundo *Cultro* 8.
demarginat,
tandem
Membrana
vel *Corio* 9.
vestit, efformat,
& affigit
Uncinulos. 10.

O 2 *Liber.*

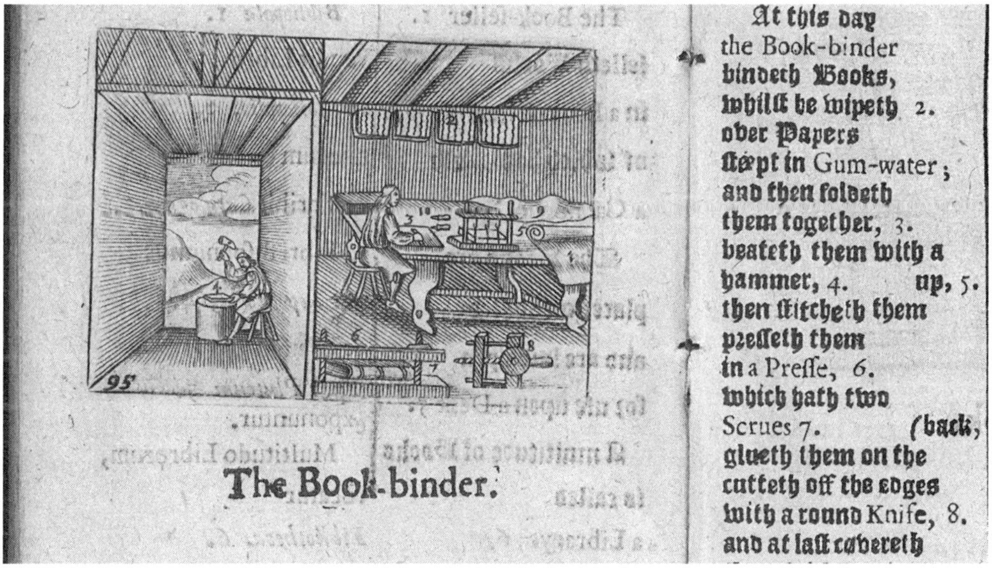

The Book-binder.

At this day
the Book-binder
bindeth Books,
whilst he wipeth 2.
over Papers
steept in Gum-water;
and then foldeth
them together, 3.
beateth them with a
hammer, 4. up, 5.
then stitcheth them
presseth them
in a Presse, 6.
which hath two
Scrues 7. (back,
glueth them on the
cutteth off the edges
with a round Knife, 8.
and at last covereth

FIGURES 25 AND 26. "The Book-binder," opening and detail from Johannes Amos Comenius, *Orbis Sensualium Pictus* (1659), C5525, sigs. O1v–O2r. Used by permission of the Folger Shakespeare Library.

Most of the books are closed, but one is open, and its pages, like the pages hanging up to dry in the bindery, have squiggly lines across them to suggest the presence of printed words. Looking more closely, we can also see smaller squiggles in the margins. The description, as translated by Hoole, explains that "within [the book] are Leaves 10. with two Pages, sometimes divided with Columns 11. & Marginal Notes. 12."

The marginal notes and the annotatibility that such notes suggest are my real concern, but the clasps and ties (or strings), included in the description of the binder's work and shown to be a nearly universal feature of the books in the last image, call attention to what N. Katherine Hayles describes as "the seriated nature of technological change."[62] Clasps were designed for books on parchment; they held the pages flat and discouraged cockling.[63] By the sixteenth century, clasps "were replaced by ties which were placed in the same positions," according to Bernard Middleton.[64] Intriguingly, the use of ties "was probably no more than a follow-on from clasps, for from the practical point of view they were not really needed."[65] Book ties are what Hayles, citing anthropologists' terminology, would call a skeuomorph, a "device that once had a functional purpose but in a successor artifact loses its functionality and is retained as a design motif or decorative element."[66] Paying attention to trends in book production that do not make economic sense (at a purely functional level), we begin to see the sociology and the ecology of the book. That is, we begin to see how printing and binding and reading habits made use of the habitats in which they arise in ornamental, not just functional ways. We could call this a poetics of making, or we could simply call it poetry, since it is an attempt to make something ornate and beautiful and more-than-functional—in this case with matter rather than words. Sizing certainly does add to the beauty of paper; type designer and letterpress printer Russell Maret talks about the "complex sensory experience" of turning crisp sheets of well-sized, handmade paper that "quiver with a gentle rattle as they are turned, making it hard to resist drumming one's fingertips against them. When bent they make a snapping sound, when shuffled they whisper like rustling leaves." But, Maret says, these are "accidental aesthetic benefits of gelatin sizing."[67] Sizing, though an added feature, was hardly a skeuomorphic, inconsequential, decorative feature of early handmade paper. It served the needs of paper users and preserved the paper for continued use.

Reading as Writing: Sinking into Similes

I want to turn now to an emphasis on paper use and consider how the presence or absence of animals on paper affected the writing experience—and especially the experience of readers writing in their books—and how those experiences are recorded not only in technical descriptions of writing, but also in what I am calling sinking paper tropes. At the outset, though, we might begin with one pervasive variety of physical evidence that has been hiding in plain sight: given the paper-surface heterogeneity discussed above, what we have long referred to as "pen trials"—those swirls and doodles that seem to be attempts to test the fitness of quill and ink—may just as likely be paper trials meant to test the fitness of the paper to receive ink. Recognizing the role of sizing, we begin to realize that we need a more detailed and more robust vocabulary to properly describe what we observe on the pages of Renaissance books. Similarly, we need to recover the vocabulary that Renaissance readers and writers used to identify the presence or absence of sizing. And although we do not find affective responses in which paper users call attention to the animals used to coat paper, we do find them using specialized language to call attention to their negotiations with the surfaces of paper. In *The Iewell House of Art and Nature* (1594), for instance, Hugh Plat offers this recipe for coating the margins of printed books so that the paper will take the marginalia: "Rvb your paper wel ouer with the fine powder or dust of Rosen and Sandrach [red arsenic sulphide] mingled in equall parts before you write therwith. . . . This is a necessarie secret for students, whereby they may note in the margentes of their bookes if the paper should happen to sinke, which is an especiall fault in many of our late yeere bookes of the Law."[68] As Plat's advice reveals, "sink" or "sinking" was a contemporary term used of paper that could not hold its ink.

The problem with poorly sized book pages, from the perspective of an end user, is that they frustrate readerly interaction. Anticipating the problem and taking the time to coat the margins of books printed on spongy paper, a reader might then add marginal notes. The homogeneity we expect of machine-made paper contrasts starkly with the heterogeneity of handmade, hand-sized paper, whose quality was affected by everything from weather patterns to the quality of plant fibers and available animal body parts. The variations in gelatin content observed by Barrett only serve to reinforce the point that writing on paper in Renaissance England was a series of ongoing negotiations—or "symbiotic relationships," to quote Helen Smith's illuminating essay on "the multitudinous uses of paper in early modern England."[69] Plat's condemnation of bad paper comes only after an exhortation, aimed particularly at students, to consider a book paper's absorbency before writing in it. Plat's language is part of the "rich esthetic vocabulary" that readers used to describe paper, and his use of the word "sinke" recalls the reading terminology that registered the experience of writing in books whose pages had varying degrees of sizing.[70] In fact, Plat's observation about the absorbency of late sixteenth-century law books is the earliest recorded usage of the now-obsolete definition of "sink": "Of paper: To cause ink to spread or 'run' on being applied to it; to absorb ink."[71]

Numerous Renaissance texts register the frustration of encountering poorly sized or unsized paper: paper that "sinks." Some references to sinking paper are technical, like Plat's description above. In Mathurin Cordier's translated Latin dialogues, which served as a crib sheet of sorts to accompany Cordier's Latin textbook, the scholarly interlocutors discuss everything from problematic ink and how to temper it so that it will not sink into writing paper, to the various uses of paper, including its use in the privy. In one dialogue, a scholar complains that he writes in earnest, but his writing is poor because he lacks "the helpes of faire writing": "Good paper, good inke, a good penne: for this my paper (as thou seest) doth sinke miserably, my inke is waterish and whitish, my pen soft and ill made." A printed marginal note glosses "sinke" as "flowe thorowe."[72] In another dialogue, a distinction is made between sinking paper and leaky ink: "Thereupon the paper* would leake." Two printed marginal notes keyed to the asterisk clarify: "*perflueret*, would sinke, or run thorow." and "*proflueret* would runne abroade, or the inke would."[73] Earnest as the conversation seems to be between the two scholars, it is hard to imagine sixteenth-century schoolboys keeping a straight face when talking about a soft pen, absorbent paper, and "waterish" ink.

We also find simple analogies between sin and sinking. In the second sermon in a printed series titled *The Saints Svbmission, and Sathans Overthrow* (1638), a series based on the command in James 4:7 to "resist the deuill, and hee will flee from you," John Preston likens the believer's heart to paper whose sizing has been compromised by moisture: "For as inke sinkes into wet paper, and runnes abroad upon it . . . so when the divell offers his snares to any empty heart, they enter in and foile him, but when the heart is fortified with the fulnesse of grace, Satans baites cannot take hold, nor enter in."[74] A more complex and materially descriptive example of the sinking paper trope appears in *Davids Hainovs Sinne* (1631). In the three-part poem comprised of 144 seven-line stanzas, Thomas Fuller retells the story of King David's adultery with Bathsheba, his failure to cover up Bathsheba's pregnancy, and his eventual murder of Bathsheba's husband, Uriah. Sent to the general bearing a letter from the king, Uriah unwittingly delivers a "speaking paper" that orders the general to put him at the front of an attack and then to abandon him.[75] In Fuller's account of David's letter:

> Then for a light, hee [David] speedily did call,
> (Thou Darknes with his project best agree'd)
> For paper, pen, and inke, to write withall,
> Though sure a poniard, might have done the deed,

Better if hee in blood had dipped it,
And on a sheet of paper what he writ,
A winding sheet far better did befit.
This certs I know, as Sepian juice did sinke
Into his spongy paper, sabling o're
The same, with various-formed specks of inke,
Which was so pure and lilly-white before:
So spots of sinne the writers soule did staine,
Whose soylie tincture did therein remaine
Till brinish teares had washt it out againe.[76]

Here Fuller imagines a range of material and metaphorical negotiations associated with the act of recording ideas on paper: pen that may as well be poniard, ink that may as well be blood, paper that may as well be winding sheet (a trope to which I return in the next chapter). We also find "Sepian juice"—the sole *OED* example of the adjectival form of sepia—and "soylie tincture" staining the whiteness of the page, which has, in the course of the extended metaphor, also become David's soul.[77] By the last line, the transformation is complete from the paper David uses to commit an atrocity to the atrocity-stained soul he will wash with "brinish teares," a reference to Psalm 51 that Renaissance readers would have easily picked up. I am most interested, of course, in Fuller's suggestion that the paper David uses to write the letter is "spongy paper" into which his ink "did sinke."[78] It is a clever metaphor that plays with the material complexities of absorption. Simon Barcham Green reminds us that "sizing is quite a subtle process; it is not just a case of being absorbent or repellent. Different uses require different standards."[79] Fuller properly understands, and, what is more crucial, believes his readers will understand, that ink really does need to sink into paper a little bit in order to render a legible message. The right amount of sponginess is good, but if paper (or one's soul) is too spongy, the ink will stain the page and render the writing illegible.

Other complex metaphors about sinking are harder to puzzle out. In Richard Harvey's *Plaine Percevall* (1590), a pamphlet against the pamphleteers on both sides of the Martin Marprelate Controversy, a six-line commendatory verse at the end of the tract creates a fairly complex metaphor in a short amount of space. Claiming that Perceval's praises will "florish in [his] dary," he says that his paper will be curds and his pen a spoon. He then claims that if his paper sinks, he will write his praises upon "a cheese, / That as the same increaseth, so may these."[80] I understand him to be likening sinking paper to cheesecloth, while also referring to the curdling of his paper/curds into cheese. But my point in citing this example is that the metaphor is convoluted and relies upon an understanding of not only the natural properties of cheese, but also the natural properties of writing on paper. A less complex, but similarly abstract metaphor occurs in the play *The Mariage Night* (1664) by Henry Falkland. The Duke claims,

> A Fair young Lady and Widow, is
> A rich piece of Stuff Rumpled: An Old one's
> A blotting Paper: A Man shall never
> write any thing on, she sinks so.[81]

In the Duke's opinion, a young widow may be pre-creased and pre-folded, so to speak, but unlike an old widow, she is still able to accept the metaphoric writing ink of a new husband. (Here, too, the language of blotting as it relates to women and chastity is complicated by the metaphor, which relies on the play between antonymic connotations of "blotting.") The chronological range of these two abstract references to animal sizing (1590 and 1664) only serves to highlight the ubiquity of bad, sinking paper.

Shakespeare's Falstaff also makes a subtle reference to the relationship between sizing and sinking. In *The Merry Wives of Windsor*, having been hidden in a laundry basket, covered with Mistress Page's "foul linen" (3.3.130), and subsequently tossed into the Thames River, Falstaff returns complaining about his supposed near-drowning experience. Lucky for him he landed in a shallow part of the river because, he says, "you may know by my size that I have a kind of alacrity in sinking" (3.5.11–12). The pun reverses material expectations: while a lack of size makes paper sink, Falstaff's ample size makes him sink. It is a subtle pun, but one that picks up on a repeated reference to paper (Mistress Page, paper letters as physical properties on stage, Falstaff's page as an intermediary, identical love letters that might as well have been printed with blanks left for the wives' names) and on Falstaff's incontinent, "leaky" intentions.

These varied references to the failure of animal sizing employed across a range of sixteenth- and seventeenth-century texts make visible an invisible substance with which Renaissance readers interacted on both a material and a metaphorical level. When annotating, they recognized the absence of sizing as a moment of sinking. When reading with their quills at rest or without pen and ink, they understood the ways in which sinking paper is likened to the sinking feeling of unrequited love, to the staining of a sinner's soul, to impregnation, or even to the technical difficulty of writing on "a cheese." Crucially, then, sizing affects book functionality and, one could argue, helps to define what is indicated by the word "reading."

There is, in addition to the sinking paper trope, another and more well-known trope that arises from the material absorbency of paper: the privy paper trope. Margaret Spufford claims that cheap print "had a secondary function of supplying the very real social need for lavatory paper," and according to Heidi Brayman, "Scatalogical uses for printed paper were often recorded and imagined in the period."[82] In one of the examples Brayman cites, William Cornwallis admits to reading ballads, but with this caveat: "My custome is to read these, and presently to make vse of them, for they lie in my priuy, and when I come thither, and haue occasion to imploy it, I read them, halfe a side at once is my ordinary, which when I haue read, I vse in that kind, that waste paper is most subiect too, but to a cleanlier profit."[83] The use described seems to be actual; at least Cornwallis believes it to be a viable excuse for reading ballads. Thomas Nashe similarly refers to the privy use of printed paper, this time of a longer printed work (rather than a ballad): "*Iack Wilton* . . . hath bequeathed for wast paper here amongst you certaine pages of his misfortunes. . . . keepe them preciously as a *priuie* token of his good will towardes you."[84] Although discussions of book survival sometimes gesture toward privy paper tropes, we tend to look to other factors, such as book format, binding, and annotation as far more predictive of a book's survival. And yet, vulgar jokes about printed pages as toilet paper are common enough to be considered indicative of widespread cultural practices. Further, what we know about sizing suggests that a reader in the privy would have a fairly easy decision to make between (1) crackly, well-sized paper that had been coated with gelatin to render it less absorbent, and (2) soft, absorbent unsized paper. Well-sized paper was necessary and useful for Renaissance readers and writers, but so was unsized paper. This observation leads to potty humor, but it is not *simply* potty humor. The correlation between unsized, sinking paper and privy use is worth further consideration and could have significant implications for the study of book survival, the topic to which I now turn.

Preservation and Survival: Animals, Archival Libraries, and Natural Selection

In addition to preparing the spongy surface of paper to accept handwriting ink, sizing also plays an important role in book survival—a role that calls into question commonly accepted beliefs about the proportional relationship between book use and book survival. Although this chapter is most focused on book use and readerly annotation, as I close I want to identify the importance of the hybrid plant/animal nature of the page in relation to book survival research models. Concerned as historical readers and writers may have been about the long-term survival of paper, especially as compared to parchment, paper records survived remarkably well for centuries before being moved to the climate-controlled microbiomes of archival libraries. On the one hand, as Ann Blair has convincingly argued, the rise of note-taking corresponds directly to the spread of paper in Europe.[85] The availability of paper encouraged manuscript note-taking and record keeping and allowed for an amplification of written records. Things that might not have otherwise been written down on a more expensive medium, grocery lists, for example, were committed to the page. The same was true for printing, as discussed in the Introduction. In addition to encouraging an amplification of original documents, the relative affordability and availability of paper also encouraged duplication (by hand and by printing press). Duplication becomes especially important as an ongoing aid to survival across the intervening centuries; Blair observes that "hardly any notes survive from antiquity: ancient papyrus had a life expectancy of two to three hundred years under normal circumstances."[86] Most of the texts that survive from antiquity survive in an altered form: few do survive in the plant form of papyrus, more in the animal form of parchment, and most in the hybrid plant/animal form of sized paper.

Paper-driven amplification and duplication aside, why do particular original items on paper—like a shopping list from 1633 found intact and legible in 2017 in the baseboards of an attic—survive so well?[87] Cathleen Baker explains that sizing "increases the strength of the sheet by 'gluing' the fibers together" and that "once sized, paper is more able to resist the absorption of liquids, pollutant gases, and airborne particulates, as well as soiling from handling."[88] In 1995, Timothy Barrett and Cynthea Mosier published "The Role of Gelatin in Paper Permanence," in the *Journal of the American Institute for Conservation*, arguing the importance of, and relative lack of research on, "the chemical and physical systems of papers made in Europe between 1400 and 1800."[89] In their study, Barrett and Mosier demonstrated that "pH, calcium content, and gelatin content . . . interact in a chemical and physical system that has an overall positive effect on the long-term stability of paper."[90] In the wake of the more recent research discussed above, Barrett and his team have further proven and more precisely defined the ways that gelatin sizing is a "net asset in paper": "The manner in which gelatin makes this positive contribution is likely to be a combination of pH buffering, added strength, preferential hydrolysis of the protein molecules over those of the cellulose (thereby slowing the acid-catalyzed hydrolysis of cellulose caused by alum), resistance to penetration of pollutant or other gasses that can contribute to deleterious reactions, and/or moderation of RH within the cellulose via a thin-film sealing effect."[91] Barrett's conclusions have important implications for papermakers and for conservators. Papermakers, who now have more modern, synthetic sizing options like Alkyl Ketene Dimer (AKD), might consider the centuries-old tradition of animal sizing. When resizing cultural heritage papers, conservators might also look to gelatin. John Baty writes, "Technologies that have this kind of track record in cultural heritage materials are great candidates for use in our work today."[92] At the heart of Barrett's research (and other research along these lines) is a deep interest in how to make and conserve papers so as to render them more lasting, more savable, more permanent.

Permanence, though, is not the sole factor in estimating the likeliness of paper survival. In 1970, in an influential essay entitled "The Nature of Paper," B. L. Browning made a crucial distinction that continues to be made by papermakers and paper conservators: "Papermakers prefer to distinguish between the concepts of durability and permanence. Durability is the property of resisting deterioration by use, that is, the handling to which paper may be subjected. . . . Permanence, on the other hand, refers to the degree to which a paper retains its original qualities during storage." Browning notes that "a paper may be durable but not permanent that is, durability may be lost rapidly because the paper contains an excessive amount of acids, or it may be permanent but not durable."[93] Sized paper is more durable by comparison with unsized paper because it can better withstand use, such as annotation. Sized paper is more permanent by comparison with unsized paper because, given identical storage conditions, the former will last longer and better than the latter. These two qualities in tandem—durability and permanence—allow us to consider that books that were made to be annotated (the most durable) are also, all conditions being equal, the most likely to survive (the most permanent). Put more directly: the more annotatable a book's pages, the more likely it is to survive.

We have much yet to learn about animal sizing and its role in making paper more durable and more permanent, but what we currently know flies in the face of what we have long believed about books being read to death. In "Playbooks and the Question of Ephemerality," Alan B. Farmer writes that "scholars have long posited that size and genre were likely to contribute to a book's ephemerality. But they have not agreed on which aspects of a book's size were most important nor which genres had higher loss rates."[94] Here, Farmer uses "size" in the dimensional sense, but the conjunction of animal sizing and genre is suggestive. For, as he demonstrates, there are high-loss genres and low-loss genres. Farmer convincingly argues that, based on the evidence, "there are genres that were ephemeral, such as catechisms, meditation and prayer books, schoolbook grammars, corantos, and pamphlets of two sheets or less," but playbooks "do not warrant being thought of as early modern ephemera."[95] If we take the role of sizing seriously as we read that claim, then we see new kinds of questions. Corantos and pamphlets had no good reason to be printed on well-sized paper and, if they were, they would not have so well served the privy needs of Cornwallis and others. But is it possible that schoolbook grammars were not well sized? And why? Cordier's Latin textbook, where students dialogue about paper absorbency, and Plat's claim about poorly sized lawbooks indicate that schoolbooks may not have been designed to accept student doodles. By contrast, given what Farmer shows about the survival of playbooks, are we to assume they were well sized? If so, the presence of sizing further calls into question what we think we know about how playbooks were read and used. Given their survival, we need not simply guess; we might use nondestructive tests to analyze gelatin content in extant volumes.

Joseph A. Dane has called into question some of the abstractions used to calculate survival and loss, and has suggested that one of the false assumptions that underlie some models is the belief that "book-copies have an equal chance of surviving." "They do not," Dane writes, and his primary example is that "some early books are printed on vellum and are thus more likely to survive as, say, binding material in other books than those printed on paper."[96] In fact, books printed on vellum might survive intact long after paper copies have disappeared: Malcolm Walsby, considering a volume printed by Jean Georget in 1541 in Rennes, France, writes that "when establishing the printed output of provincial printers it is necessary to be aware of the poor survival rate. The sole surviving copy [of Jean Bouchet's *Les triumphes de la noble et amoureuse dame*] was printed on vellum, a much tougher material than paper that would have increased the book's resistance to repeated reading and rough handling."[97] At the most basic level, then, statistical models used to calculate book survival rates should consider the significant quality difference between books printed on paper and books printed on vellum. Considering a binary like vellum versus paper is better than ignoring the effects of substrate quality upon survival, but it is more a starting point for suggesting that survival statistics that do not account for substrate quality have innate limitations that we might begin to address.

As this chapter has suggested, our investigation into substrate qualities and their impact on survival could offer more granularity to a branch of book investigation that seems to thrive on granularity. "Questions of survival and loss bedevil the study of early printed books," claims the publisher's description of *Lost Books: Reconstructing the Print World of Pre-Industrial Europe*, an impressive volume filled with over five hundred pages of research by leading specialists.[98] While the survival calculations cited in *Lost Books* correct for numerous variables, substrate quality is not one of those variables. The heftiest, most recent, most broadly conceived work of scholarship on the survival of early books mentions vellum once (but in relation to manuscript books); "gelatin" and "sizing" never appear, and when "size" does appear, it is in reference to a volume's dimensions—not the strength and quality of its pages. I see this as a tremendous opportunity for new collaborative research across disciplinary boundaries, research that is attuned to the ecology of the media in ways that generate more precise, more nuanced, and more holistic statistical models for understanding book use and survival. New approaches to accounting for substrate quality need to be considered, examined, and tested with available technologies if we truly believe that, as Alexandra Hill asserts in "Lost Print in England: Entries in the Stationers' Company Register, 1557–1640," "by ignoring lost works, we can never hope to gain a full picture of early modern printing in England."[99]

Filled as this chapter is with references to what is missing, even in the most impressive scholarship in the field of book history, the work here is intended as a formal acknowledgment of an elephant in the room—or rather, of the herds of domesticated animals in the sizing room—that are so easy to overlook. At present, scholarship on sizing tends to focus on questions of production and, especially, of preservation and conservation, but ongoing research on gelatin (and alum, and calcium, etc.) in early handmade papers has the potential to disrupt what we think we know about book use and book survival. In a very simple narrative we sometimes retell about the history of the book, plant gives way to animal gives way to plant media as records move from papyrus to parchment to paper. In reality, at least in Western paper, plant gives way to animal gives way to plant-animal hybrid. When Hamlet recites lines that describe Pyrrhus as "o'ersizèd with coagulate gore" (2.2.487), he uses a phrase that also aptly describes early handmade paper, or at least the sort of paper that Hamlet would need in order to "set down" a speech of "some dozen or sixteen lines" on paper (2.2.567–68). Recognizing—re-seeing—animal sizing, we begin to understand and learn from the complex ways in which reading habits and reading habitats are always systemically entwined. Seeing animal presence in the plant-based pages of early printed books may lead us to think more carefully about Renaissance ecology or ecocriticism or categorical hybridity or media consumption, but it also prompts material discoveries about books we now explore almost exclusively in climate-controlled microbiomes designed to keep decay from disfiguring the fluctuant records we have gathered from the past. It is to concerns about the environmental costs of preservation, and to the ecopoetics that arise from such concerns, that I turn in the final chapter.

CHAPTER 5

This Book, as Long Lived as the Elements: Climate Control, Biodeterioration, and the Poetics of Decay

> Not only marble, but the plastic toys
> From cornflake packets will outlive this rhyme:
> I can't immortalize you, love—our joys
> Will lie unnoticed in the vault of time.
> —Wendy Cope, "From Strugnell's Sonnets"

In Chapter 2, I considered Henry Vaughan's "The Book" as a poem that imagines the various plants and animals used in the making of an "aged book"—probably an old Bible. Although Vaughan's attention to the ecology of his media is exceptional, the circumstances of the poem's writing are familiar. When we consider Vaughan looking at an antiquated book and imagining how it came to be, we realize that his basic approach was not unlike the approach of present-day scholars in archival libraries: like Vaughan, we talk about a book's history, about how it has been composed. We, like Vaughan, try to reconstruct backstories from traces left in book objects: linguistic evidence that reveals which manuscript and print sources a writer drew upon while writing; typographical and paleographical anomalies that suggest how and when a book was made; manuscript annotations that indicate the reactions of readers. We might even observe, especially when scanning archival catalogues, how the careless hands of users and of time have changed book objects, leaving them torn, incomplete, blotted, burned, wrinkled, worm-eaten, brittle, faded, and foxed. Almost without exception, we look for stories from the past in old books.

Unlike Vaughan, however, we tend not to think about a book's inevitable decomposition at some point in the future. We expect, quite reasonably, that decomposition is being kept at bay because, unlike Vaughan, we usually look at sixteenth- and seventeenth-century books from within climate-controlled, manufactured environments where temperature and relative humidity are carefully monitored and regulated. Reading Vaughan, Chapter 2 examined an ecological history—even an etiology—of paper. There, like Vaughan in "The Book," I start with the thing itself in our hands and ask, "How did this book come to be?" and "How did readers imagine how the book came to be?" With Donne and with Shakespeare, this chapter thinks forward to preservation, to questions about how a book ceases to be. This chapter is guided by two questions: "How does a Renaissance book survive?" and "How does one imagine the survival of a Renaissance book?" The first question is more technical, and the second more poetical, but, as I demonstrate, the technical and poetical entangle and overlap in intriguing ways, and these entanglements become more profound, and more profoundly poetic, when we reframe these two questions in terms of decay rather than survival: "How does a Renaissance book decay?" and "How does one imagine the decay of a Renaissance book?"

Central to this chapter, then, is a concern with how ecological conditions of book preservation inflect reading practices. And while the ecological conditions of book preservation certainly affect readers (in the sense that they stimulate affective responses) and inform readers (in the sense that they allow readers access to the past), I am equally interested in how the environmental conditions that allow us to handle old books distract and mislead us, how they trick us into imagining that survival is normal and that decay or loss is notable. "The notion of a perfect, stand-alone copy reflected the system of industrial book production that was standard in modernity," writes Jeffrey Todd Knight of the ideal of a "collated and perfect" book, and it "also reflected a system of economic value that determined the fate of books inherited from earlier eras of production."[1] We see this system of back formation reflected in rare book catalogues, where "imperfect" copies with ripped pages, missing pages, blots, and burns are listed as if they are exceptional.[2] Hiding the nature of the book and its tendency toward decay, don't we also hide the very quality that enlivens so much Renaissance poetry? The corruptible matter used by Renaissance poets to record ideas for posterity both threatens and inspires poetic expression, for the very ecosystem that supplies the materials needed to give life to poetry on paper also thrives on death and decay. In archival libraries, books are carefully monitored to be assemblages of matter in an arrested—or, at least, a delayed—state of decay. They, or we, or both are time travelers, interacting with untimely ecologies, inspecting the not-quite-corrupt remains of long-expired plants and animals.

Here and there, rumblings of what could become literary-critical conversations about archival library environments in the Anthropocene can be heard mostly in the work of medieval scholars. Siân Echard, summoning "Jerome McGann's appeal that we use bibliography to give our texts 'a local habitation and a name,'" has drawn attention to "questions of how archival practices and archival encounters structure and control our reading of medieval books and the texts they contain."[3] Jennifer Summit, taking up these questions in *Memory's Library: Medieval Books in Early Modern England*, explores the layered contexts that control archival access. According to Summit, the question "How did this book, so far from its original place of production and reception, reach me in the present?" is one that "today's readers of medieval books"—and, we could add, readers of early modern books or any books, really—"must continually ask themselves."[4] When we do question how a book got from there to here, how it was, in Summit's words "transport[ed] across time and place, from monastery to well-lighted and guarded modern reading room," we typically discuss a book's survival in terms of human provenance.[5] Even Echard's and Summit's questions, which come closest to acknowledging the importance of the archival library environment, highlight human affect more than human ecology.[6] In our fascination with the ways that human use changes text, it is easy for literary critics to forget that the vaults of archival libraries guard books not only against human damage and theft, but also against nonhuman damage and theft such as biodeterioration (a word I discuss in more detail below). As Henry Vaughan's Bible must have warned him, unless the page from the Gospel of Matthew was missing or damaged, moths and rust are bound to corrupt earthly treasures.[7]

Ecocritics, no less than bibliographers, overlook the importance of the archival library environment or microbiome, the constructed geographic and climatic zone in which many of the books we study are saved from extinction. For example, Simon Estok claims that "Shakespearean ecocritics still seem to lack the professional dignity of, say, the archivists who burrow in the bowels of the Folger."[8] But he is being dramatic, or possibly playful. A simple binary between ecocritic and archival library researcher does not hold, for these collections are actively used by literary scholars of all critical persuasions. More notably, the concerns about scholarly dignity upstage environmental precision: scholars do not "burrow in the bowels of the Folger"—it is simply too cold to work comfortably down there for long periods of time, as Estok must know.[9] Even on short excursions to the lower decks one may find scholars trying to carry books while simultaneously warming hands in armpits. Upstairs, outside observers looking in at researchers in the reading room during a Folger Shakespeare Library tour might think we are more appropriately dressed for a snowball fight than for research. The great majority of scholarly researchers and writers access at least some of the texts that make scholarly output possible from within climate-controlled microbiomes where temperature and relative humidity levels are making the books we hold less biologically dynamic than ever before.

Although humanities scholars may not be actively considering the relationship between local archival library environments and the global environment, those charged with the task of conservation certainly are. It is in their scholarship that we find not only active concern with the future of rare books, but also active, collective attempts to imagine what the environmental future of these books may be. Since 1997, the Getty Conservation Institute (GCI) has been investigating "economical and sustainable climate improvement strategies (i.e., effective alternatives to air-conditioning systems)" to "preserve collections in regions characterized by hot and humid climates."[10] In 2007, "based on the growing need to adapt to global climate change, along with a call for more sustainable strategies," GCI hosted an "interdisciplinary dialogue about the museum environment, specifically the indoor climate" with "an international group of experts—including architects, engineers, conservators, and conservation scientists."[11] As part of the dialogue, Ernest A. Conrad, an engineer who was involved with the design of climate control systems for other major cultural institutions in Washington, DC, such as the National Gallery of Art and the Library of Congress, explored the troubling, future-facing question: "What if a 1°C global temperature increase should occur? How would that increase affect the roles of those of us working in the profession of preservation?"[12] Although Conrad notes that "well-financed" museums could survive such an increase without significant impact, he pointed out that the bigger concern is that "extreme weather changes caused by global warming are the most serious concern for all institutions, no matter where in the world they are located."[13] Another participant, former UK National Trust Historic Properties director and head conservator Sarah Staniforth, subsequently revised and published her remarks as "Slow Conservation," a *Studies in Conservation* article that advocated for a conservation movement in the tradition of the Slow Food movement. "Our challenge in the twenty-first century is to achieve appropriate preventive conservation standards in a way that does not use excessive energy," writes Staniforth, observing that "the reality of climate change has added impetus in the twilight years of carbon-based fuels, to make energy-use reduction a driver for all organizations."[14] Hopeful as these new preservation practices are, the truth behind them, sometimes named and sometimes pushed to the periphery, is that the question, as it relates to the catastrophic loss of millennia of human records, is not "What if?" but "When?"

The hope of Shakespeare's *Sonnet* 81 is a fantasy, it would seem:

Or shall I live your epitaph to make
Or you survive when I in earth am rotten.
. . . .
Your name from hence immortal life shall have,
Though I, once gone, to all the world must die.
. . . .
Your monument shall be my gentle verse,
Which eyes not yet created shall o'erread;
And tongues to be your being shall rehearse,
When all the breathers of this world are dead. (81.1–2, 5–6, 9–12)

Here the rotting of the poet's body and the breath of his contemporary readers are juxtaposed with "eyes not yet created" and "tongues to be" who will see and read a monument made with paper. Decay is embedded in the conversation about survival. Poetry and preservation are part of the same imaginative task. Imaginative, but also actual and effective: four hundred years later, we are the eyes and tongues the poet imagines. Toward the end of this chapter, I return to Shakespeare's *Sonnets* and consider more carefully how words on paper have been, at least until now, immortalized and "rehearse[d]" (a word that calls attention to a dead body, to memorialization, to reading aloud, and to decay and decomposition all at once). Here at the outset, I want to emphasize the places and times in which these rehearsals happen, places and times that become impossible to overlook if we take seriously (1) that books are composed of once-living organisms and that those once-living organisms are now actively, if very slowly, decomposing; and, (2) that current strategies for stalling book decay are not sustainable. Early in this chapter, I focus on a book that John Donne imagines lasting as long "as the elements, / Or as the world's form."[15] In the poem, Donne suggests that the book will somehow evade quotidian decay and destruction—which is to say, the most obvious threats to a Renaissance book's survival—and will be destroyed only by a global catastrophe.[16] In an ironic turn, four centuries later we find that threats to a Renaissance book's survival are neatly reversed: Donne's legacy in the forms of paper records, which are well preserved against environmental decay in climate-controlled vaults, are more likely to be destroyed by a sixth extinction than by attrition.[17] In this sense, whether we acknowledge it or not, Donne's concerns about the decay and survival and preservation of records are also ours.

Like the poets this chapter considers, I have been using the word "decay" to indicate the natural process of paper deterioration, but the word "biodeterioration" more precisely indicates the codependent life (bio-) and death (deterioration) interactions that take place when a text decays. The *OED Online* defines "biodeterioration" as the "deterioration of a substance or object caused by the action of living organisms," but H. J. Hueck, who is credited with coining the term, offers a definition of "biodeterioration" that, in its surprising emphasis on a subjective point of view, is much better suited to this conversation: biodeterioration is "any undesirable change in the properties of a material caused by the vital activities of organisms."[18] If a scientific definition can qualify as poignant, then Hueck's does. On the one hand, the deterioration is just a part of life—even part of life giving. It is one of the "vital activities" that sustain organisms. On the other hand, those same vital activities, as Hueck's definition recognizes, effect change that is "undesirable" to someone, somewhere. We may be inclined to encourage the biodeteriorating activities that cause, for instance, apple peels and coffee grounds in a compost heap to become rich additives for gardens (or rags to break down to cellulose so they can be made into paper); however, we are less enthusiastic about the biodeterioration of, say, an extant Donne manuscript or a first edition of Shakespeare's *Sonnets.*

I claimed at the outset that this chapter is driven by questions of how books *actually* decay (or survive) and how books are *imagined* to decay (or survive). But while drawing on scientific language—like biodeterioration and microbiome—that allows a more precise, complex conversation about what is happening to books inside archival libraries, this chapter most engages with the latter of those two questions, that is, with how books are imagined to decay. Believing it is the collaborative work of scientists and of humanities scholars to draw attention to and find creative, intentional solutions for dealing with our rapidly changing planet, my work here takes up the task of calling attention to an inconvenient truth: there is an inverse correlation between book preservation strategies and environmental stewardship. The preservation of records has a significant environmental cost, and that cost is expected to rise precipitously as resources become scarcer. Jonathan Bate has written that "the price of art is the destruction of a living tree. . . . You create culture by enslaving nature."[19] We might say that the current price of art preservation with HVAC systems is the destruction of a fossilized tree; we must continue to enslave nature in order to preserve the culture we have created.

Donne's Elemental Book

In Vaughan's "The Book" a very much alive poet is looking at a very much dead collation of "scatter'd" flora and fauna in the form of a leather-bound volume printed on paper. He imagines its past and its future, its composition and its decomposition, but his meditation is grounded in the present and focused on a "dry and clean" volume that does not yet seem to be an actively decaying object.[20] It is an "aged book," but not one with papers "yellowed with their age," as in Shakespeare's *Sonnets* (17.9) or "worme-eaten and full of canker holes," as are the records in Eumnestes's library in *The Faerie Queene*.[21] Vaughan's "The Book," a meditation inspired by an intact book, is a close reading that looks backward at the book materials that have been used, how they have been used, and how they may be reused.

By contrast, John Donne's "A Valediction: Of the Book" is a poem about an imaginary book that does not yet exist. This imagined book, the poet suggests, would be made from extant "manuscripts" and "myriads / Of letters, which have past 'twixt" himself and the beloved of his poem, presumably his wife, Ann. At the outset of the poem, the poet makes clear that the goal of this imagined book is to cheat the common destiny of death, to allow poet and beloved to "stay" and "out-endure" after they are gone so that "posterity shall know" about them. The book object has value, in Donne's poem, as a memorial container for his (and his love's) words. The opening lines of the poem themselves seem to be recording for us, the listening or reading audience, a snippet of a private moment from the middle of a conversation between the two lovers:

> I'll tell thee now (dear love) what thou shalt do
>   To anger destiny, as she doth us,
>   How I shall stay, though she esloign me thus,
> And how posterity shall know it too;
>   How thine may out-endure
>   Sibyl's glory (ll. 1–6)

The poem is subtly crafted—filled with internal rhymes and irony—and yet, by the seventh and last stanza, it ends up sounding not unlike a book prospectus.[22] The first stanza states the poet's goal, and the second outlines practical considerations. The volume's primary archive will be extant love letters. According to the second stanza, the book's intended audience will be those "whom love's subliming fire invades" (it will have cross-disciplinary appeal) (l. 13). The book will be actively endorsed by Love, who has given them special permission to speak on his behalf. The volume will not only record the love of the poet and his beloved, but may be adopted as a textbook of sorts where students of love will find both "Rule and example" (l. 14). Skipping over stanza 3 for a moment, we find that stanzas 4 through 6 paint a more detailed image of the volume's potential readership ("Love's divines" in stanza 4, "lawyers" in stanza 5, and "statesmen" in stanza 6) before stanza 7 launches from material considerations into a closing metaphor (ll. 28, 37, 46).

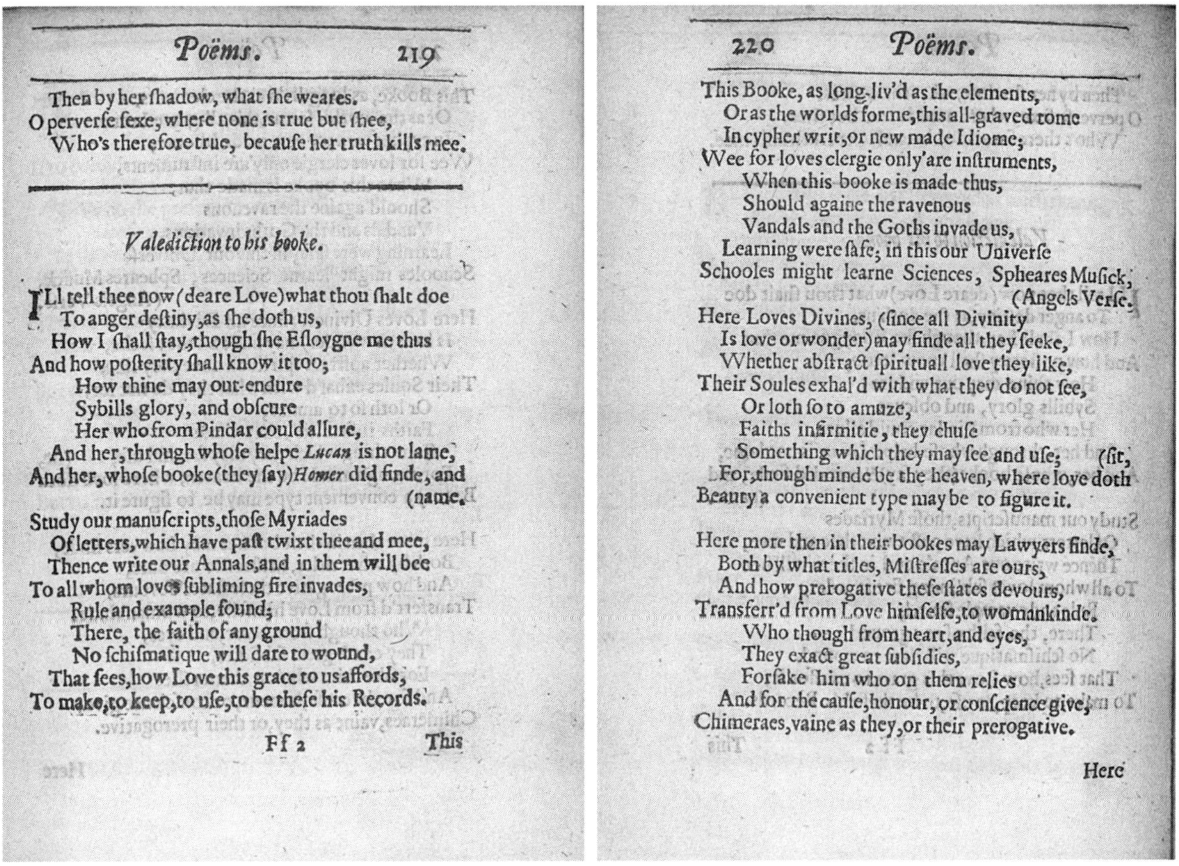

Poëms. 219

Then by her ſhadow, what ſhe weares.
O perverſe ſexe, where none is true but ſhee,
Who's therefore true, becauſe her truth kills mee.

Valediction to his booke.

I'Ll tell thee now (deare Love) what thou ſhalt doe
To anger deſtiny, as ſhe doth us,
How I ſhall ſtay, though ſhe Eſloygne me thus
And how poſterity ſhall know it too;
How thine may out-endure
Sybills glory, and obſcure
Her who from Pindar could allure,
And her, through whoſe helpe *Lucan* is not lame,
And her, whoſe booke (they ſay) *Homer* did finde, and
(name.
Study our manuſcripts, thoſe Myriades
Of letters, which have paſt twixt thee and mee,
Thence write our Annals, and in them will bee
To all whom loves ſubliming fire invades,
Rule and example found;
There, the faith of any ground
No ſchiſmatique will dare to wound,
That ſees, how Love this grace to us affords,
To make, to keep, to uſe, to be theſe his Records.

Ff 2 This

220 Poëms.

This Booke, as long-liv'd as the elements,
Or as the worlds forme, this all-graved tome
In cypher writ, or new made Idiome;
Wee for loves clergie only'are inſtruments,
When this booke is made thus,
Should againe the ravenous
Vandals and the Goths invade us,
Learning were ſafe; in this our Univerſe
Schooles might learne Sciences, Spheares Muſick,
(Angels Verſe.
Here Loves Divines, (ſince all Divinity
Is love or wonder) may finde all they ſeeke,
Whether abſtract ſpirituall love they like,
Their Soules exhal'd with what they do not ſee,
Or loth ſo to amuze,
Faiths infirmitie, they chuſe
Something which they may ſee and uſe;
(ſit,
For, though minde be the heaven, where love doth
Beauty a convenient type may be to figure it.
Here more then in their bookes may Lawyers finde,
Both by what titles, Miſtreſſes are ours,
And how prerogative theſe ſtates devours,
Transferr'd from Love himſelfe, to womankinde.
Who though from heart, and eyes,
They exact great ſubſidies,
Forſake him who on them relies
And for the cauſe, honour, or conſcience give,
Chimeraes, vaine as they, or their prerogative.

Here

FIGURE 27. John Donne, "Valediction to his booke," in John Donne, *Poems, with elegies on the author's death* (1633). Courtesy of the Department of Special Collections, Memorial Library, University of Wisconsin–Madison.

The energy of the poem centers (and, some have argued, peaks) at stanza 3.[23] It is in this stanza that the poem takes a dramatic turn from the first two stanzas' preoccupation with the imaginary book's creation to a touching consideration of the imaginary book's eventual destruction, an event that, the poet imagines, will happen only when the elemental world is destroyed. Donne goes on in that third stanza to imagine a book that will have the power to persist and to preserve knowledge even in the event of violent human destruction:

This book, as long lived as the elements,
Or as the world's form, this all-graved tome
In cypher write, or new made idiom;

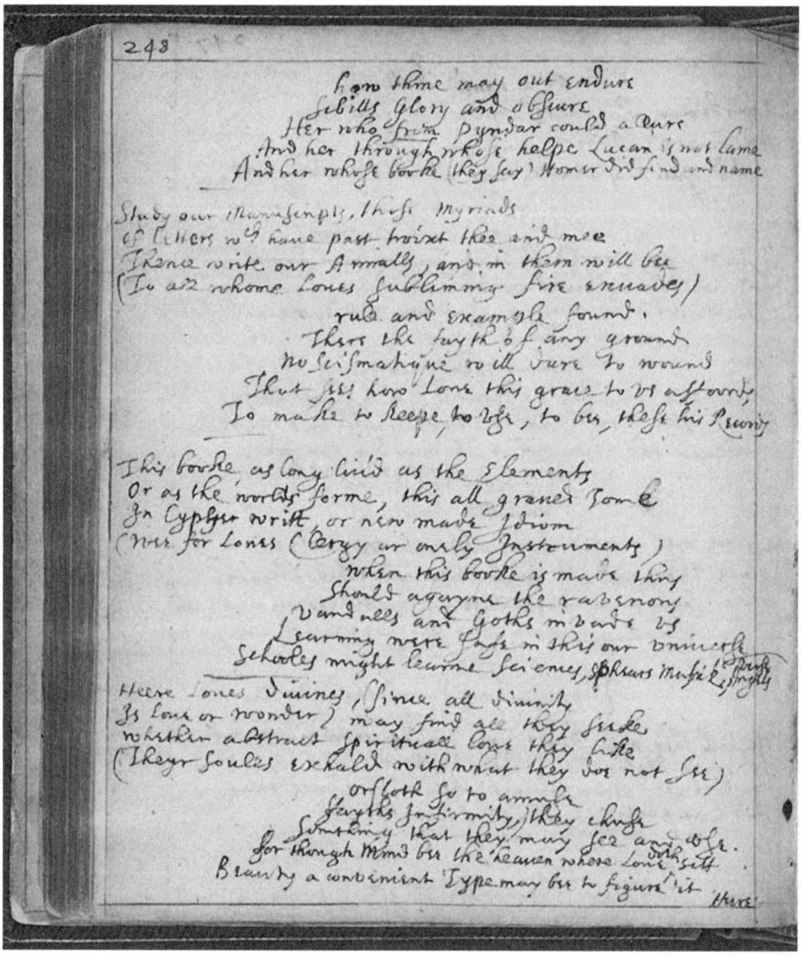

FIGURE 28. John Donne, "Valediction of the booke," in the O'Flahertie Manuscript, MS Eng 966.5, Houghton Library, Harvard University.

We for Love's clergy only are instruments,
When this book is made thus,
Should again the ravenous
Vandals and Goths inundate us,
Learning were safe; in this our universe
Schools might learn sciences, spheres music, angels verse. (ll. 19–27)

Like so much of Donne's poetry, "A Valediction: Of the Book" (hereafter, I use "Of the Book" when referring to the poem) engages not only with the complexities of recording ideas on matter, but also with the challenges of creating permanent memorials for impermanent people, of immortalizing mortality on mutable objects that tend toward decay. Peter McCullough has claimed that "death and mortality were, throughout [Donne's] entire writing life, what seemed most to make him reach for his pen."[24] He is a writer fascinated with absence and with presence, with corruption and with preservation. It is no surprise, then, that we find in Donne's work, from early lyrics and satires to his sermons, a preoccupation with written language as a kind of preservatory, an embodiment, a reanimating object.

Donne's reflections on the ecology of the book and the longevity of his love lyrics and elegiac poetry engage with a wholly different set of concerns than Vaughan's meditations on the ecology of his Bible in "The Book." While Vaughan's poem highlights the ecological history of a volume, Donne's imagines a volume's ecological future.[25] In "Of the Book," Donne focuses entirely on an imagined, future book and on concerns about the human record itself, about how long it will survive, how it might outlast marauding Vandals and Goths, how it might be read and understood by posterity. When we hold a four-hundred-year-old book in which a poet worries about how their words will be decayed, it is easy to imagine such lines as performative, even clichéd protestations. Shakespeare's *Sonnets,* for instance, have been so actively reproduced in the centuries between their publication and now that it is hard to take seriously lines such as these:

> But wherefore do not you a mightier way
> Make war upon this bloody tyrant Time,
> And fortify yourself in your decay
> With means more blessèd than my barren rhyme? (16.1–4)

The idea that Shakespeare's *Sonnets* were barren is laughable, and the sentiment now reeks of excessively false modesty. In *Sonnet* 65, the poet suggests the preservation of his paper memorial would have to be miraculous to succeed. How can his poetry survive, he wonders,

> Since brass, nor stone, nor earth, nor boundless sea
> But sad mortality o'ersways their power,
> How with this rage shall beauty hold a plea,
> Whose action is no stronger than a flower?
> . . . .
> When rocks impregnable are not so stout
> Nor gates of steel so strong, but Time decays?
> . . . .

> Or who his [Time's] spoil of beauty can forbid?
> O, none, unless this miracle have might,
>
> That in black ink my love may still shine bright. (65.1–4, 7–8, 12–14)

Though there may well be a degree of false humility in the trope of decay to which poets often turn, there is also a degree of logical anxiety about how to make embodied words "out-endure," to borrow Donne's word, the decay of the human hands that wrote them. For Donne, the inevitable deterioration of human-recorded words over time generates poetic conflict—and, significantly, the threat of loss generates poetic matter.[26] (And not just for Donne; as I have argued elsewhere, in his desire to thwart time, the speaker of Shakespeare's *Sonnets* adopts a strategy suited to an ecosystem in which time actively decays poetry.[27])

The question that breaks unexpectedly into the third stanza of Donne's bookish meditation is worth taking perfectly seriously: How does a mortal poet like Donne use a corruptible media object like handmade paper to create a memory of his mortal beloved that can be read and discussed and considered four hundred years later? What are the conditions that allow corruptible organic objects to carry messages into the distant future? Or, to return to *Sonnet* 65, what "miracle" can save paper poetry from the ravages of Time in a world where brass and stone, "rocks impregnable" and "gates of steel," are helpless against decay. Answer: air-conditioning.

Outside: Rock, Paper, Spores

The inside of the climate-controlled Folger Shakespeare Library may be warding off decay, but the outside confirms Shakespeare's belief that even "rocks impregnable" cannot win against Time. There, on the facade of the library, the stone forms of Romeo and Juliet are being threatened by biodeterioration. The exterior of the library, clad in white Georgia marble (the same sort of stone used for the statue of Lincoln in the Lincoln Memorial) and ornamented with Shakespeare bas-reliefs that sculptor John Gregory completed in the 1930s, faces biodeterioration problems that the interior does not. Inside, nearly a third of all extant copies of the First Folios evade decay and destruction, giving new meaning to Ben Jonson's famous claim, engraved on the wall above the bas-reliefs: "Thou art a Moniment, without a tombe, / And art aliue still, while thy Booke doth liue."[28] At the Folger, as at archives and libraries around the world, paper monuments from 1623 fare better than stone monuments carved three centuries later, even stone monuments carved in what the Georgia Marble Company was advertising ca. 1927 as a medium "unexcelled for sculpture work exposed to the weather, the marble being unaffected by even the most severe weather."[29] Its acclaimed durability notwithstanding, the marble book that Hamlet holds in his left hand is crumbling, and two of literature's most well-known lovers are being wrenched apart all over again. In comparison to the other nine marble reliefs, Romeo and Juliet have withstood erosion quite well, but decay has settled into the scene in the form of gypsum staining and biological spores (see Figure 29). The growth is especially heavy between the young lovers' almost-touching lips and between their interlaced fingers. Here, the marble surface has worn down, allowing biological spores to take root in the porous cavities of the stone. Having taken root, and without treatment, these spores will multiply. The words "kill the growth" appear in the conservator's assessment.[30] Growth, of course, being what Hueck would call a (humanly) "undesirable change . . . caused by the vital activities" of spores. And yet it is as natural for spores to desire the cavities of marble as it is for teenagers to desire each other. (And, in this case, as deadly.) At the level of spores and stone, conservators are the unyielding parents fighting to keep the undesirable but vital activities of procreation at bay.

The paper objects inside that are deteriorating less rapidly than the marble outside call to mind Shakespeare's bold prediction, in *Sonnet* 55, that "not marble nor the gilded monuments / Of princes shall outlive this powerful rhyme" (55.1–2). Inside the cracked, gypsum-crusted, eroded, spore-colonized, stained walls of white marble, there are two extant copies of Shakespeare's *Sonnets* printed in 1609. Shakespeare would have no reason to believe that the pages of a first edition of his volume would feel sturdy, flexible, and clean in the hands of someone living four centuries in the future. All evidence that Shakespeare would have seen was that paper books, especially small, slim volumes, had a better chance of being destroyed or repurposed or burned or dropped in the water or otherwise mutilated than of being preserved. If the paper were left outside, it would not feel so hardy. Exposed to the same conditions as the marble bas-reliefs, in Washington, DC's heat and humidity (never mind the rain), the plant-based paper would have disintegrated and disappeared long ago. In a humid climate, especially, more "fungi constitute a major threat to the preservation of archival material compos[ed] of paper."[31] One team of researchers studying the risks of fungal spoilage in paper archives note that the organic nature of paper makes it susceptible to over two hundred fungal species, which are "the main cause of damage to objects of cultural heritage made of or supported on paper."[32]

Looking back from the present, we know the long-term effects of biodeterioration on paper archives. We know that Renaissance books have a high loss

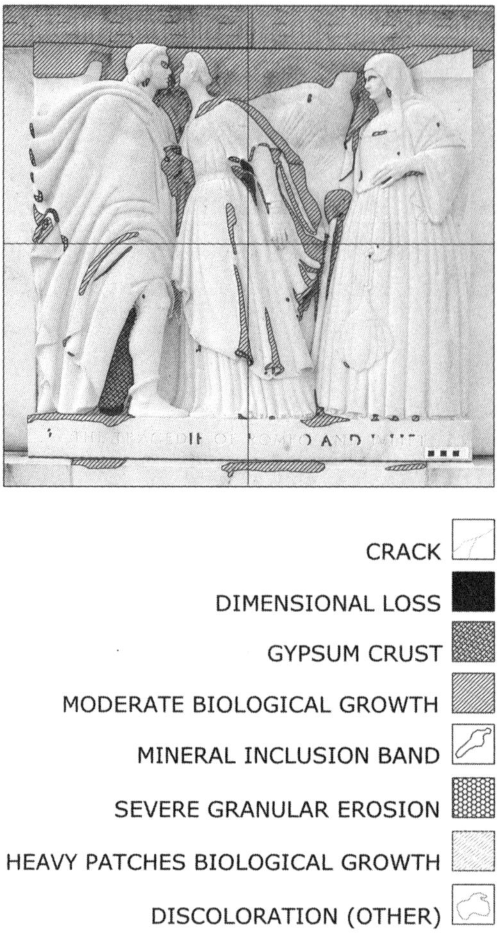

FIGURE 29. *Romeo and Juliet* (bas-relief), conservation assessment in Guy Munsch, "Condition Report for the Folger Shakespeare Library, July 2002." Used by permission of the Folger Shakespeare Library.

rate.[33] And we know that those "that survive in large numbers do so essentially because they were destined for libraries."[34] But we sometimes overlook what Renaissance readers and writers might have known—or not known—looking back from the 1600s at the history of paper records in England. As previous chapters have discussed in detail, paper did not spread into England until the fourteenth century (and papermaking was not established until the end of the fifteenth century—and even then, papermaking had more of a toehold in England than a foothold). The oldest paper that Shakespeare could have seen would have been roughly half the age of the paper we now leaf through when we read Renaissance books in the archival libraries. He had no way of knowing how long a poem on paper could last. He had no way of knowing, as Alan Galey argues in a brilliant reading of a misquotation from *The Tempest* engraved on the research entrance of the National Archives Building, that even though his manuscripts have not survived, he "lends authenticity to the idea of archiving itself."[35] More crucially, Shakespeare and most of his contemporaries had no way of guessing that such paper might be preserved by making a room cold.

In *The New Organon*, considering where streams of experimentation, like papermaking, might lead, Francis Bacon highlights the importance of better understanding "the nature and effect of cold," a topic that proved to be a singularly problematic line of inquiry for Bacon.[36] Discussing the effects of heat and cold and human experimentation, Bacon admits that "human power is plainly lame in one foot. We have the heat of fire. . . . But we lack cold, except what can be had in winter or in caves or by packing things in snow and ice."[37] He admonishes that "we must use all diligence to gather instances of cold," and since "nature supplies cold so infrequently . . . we must carefully inquire whether there are any substitutes for cold."[38] His own diligence in this direction apparently led to his death in April 1626. As James Spedding explains, Bacon

> took advantage of an unseasonable fall of snow to try whether it would preserve flesh from putrefaction, as salt does; and so caught the chill of which he died. The story is, that the idea suddenly occurring to him, he stopped the coach, alighted at a cottage, obtained a hen, helped to stuff it with snow with his own hands (and I suppose carried it with him to watch the effect); but on the way found himself suddenly so ill that he was forced to take refuge in a house of Lord Arundel's.[39]

In Bacon's own words—dictated because his fingers were "so disjointed with this fit of sickness" that he could not "steadily hold a pen"—he had tried "an experiment or two, touching the conservation and induration of bodies."[40] In a bit of irony, the secret of refrigeration that eluded but killed Bacon now sometimes causes those who research Bacon and his contemporaries in archival libraries to worry aloud about freezing to death.

Having halted or at least slowed decay, our climate-controlled "archival temples" seem to enable a fantasy of textual fixity and poetic uniformity that is as imaginative as the fantastically durable volume that Donne imagines in "Of the Book."[41] Vandals and Goths should be the least of our preservation concerns as we face the grim but near-certain prospect of global climate crisis. But even if we choose to ignore the warnings of climate disaster, it is worth considering what we overlook when we imagine climate-controlled fixity as the normal state of being for a book and for the varied ecologies from which a book is made. Imagining a lyrical poem on paper as a stable object capable of being preserved—essentially, refrigerated—indefinitely in an archival library, we overlook its poignancy. We tend to talk about the afterlife of the books we examine; Renaissance poets tended to fixate on the "afterdeath" of their recorded words.

Fear of loss and decay is one great inspiration of Renaissance poetry. A slow death, and possibly resurrection in new book objects, were perhaps the most hopeful legacy Renaissance poets could imagine for their poetic progeny. This is not to say that all Renaissance poets fully believed that their words would disappear, nor do I mean to suggest that all paper-based archives not buried in climate-controlled vaults would otherwise be lost, but given what Renaissance poets knew about books and about the plant-based materials on which their words would be circulated, they expressed valid material concerns about preservation.[42] The idea that an intact, author-approved text would be read centuries later was, like the idea of a flea commingling the blood of two lovers, the fanciful stuff of Donnean conceits.[43]

"Man-Made Weather" and Microbiomes

What we begin to realize, when we question our own access to aging textual materials, is precisely that which Bacon suspected when he so ardently advocated for further inquiry into "substitutes for cold": human-engineered cold has a profound effect upon human society and on the preservation of everything from personal comfort to dead chickens to media objects. In fact, the making and preserving of media have more than an anecdotal connection with HVAC systems and air-conditioning. In 1902, "what is generally acknowledged to be the world's first air-conditioning system" was installed in Sackett-Wilhelms Lithographic Publishing Company in Brooklyn, New York, in an attempt to make multicolor printing viable.[44] According to Gail Cooper, the problem that the publishing company faced was that "fluctuating levels of humidity caused the paper to swell or shrink, effectively changing the size of the paper each time it passed through the press" and causing "the failure of the colors to register in the proper position."[45] Two decades after his invention was first installed in a printing house, in a 1933 *Scientific American* article titled "The Economics of Man-Made Weather," Willis Carrier claimed that "more than 200 industries have already found air conditioning an indispensable servant, freeing the progressive manufacturer from daily weather uncertainties, improving the quality of his product, and contributing to the health and efficiency of his workers."[46] The article goes on to predict that air-conditioning could revolutionize home comfort as well.

History has borne out Carrier's prediction about home comfort: a recent *New York Times* article reports that the technology that made it possible for cities like Phoenix, for instance, to grow from 100,000 people in 1950 to 1.5

million in 2017 has now spread north to the extent that "eighty-six percent of new single-family homes in the Northeast [United States] are now built" with centralized air-conditioning.[47] Indoors, cooking or reading or sleeping in a climate-controlled home, many of us now rely on media to tell us how hot and/or humid the air is on the other side of the wall. When the Folger was built in 1932, only the vaults had air-conditioning, and Henry Clay Folger naturally assumed no one would use the reading room in August; now August is one of the reading room's busiest months. Walking out of the heat and humidity into an archival library like the Folger or the Newberry—or like the British Library, which is less than four miles from Highgate, where Bacon died—we enter an environment that Bacon's snow-stuffed chicken anticipated, but one that even Bacon could not have imagined when he was conducting his fatal experiment on induration. What are we supposed to call the environment of the British Library? It is London, but it could be anywhere. Or nowhere. It is easy to imagine the preservation of books as an ungrounded activity that just happens, as a system or communications circuit with no environmental inputs, as a sort of utopia where natural resources are unlimited. But books are made in and from an environment, and they are preserved in and from an environment. They come into being somewhere and they will cease to exist somewhere. Thinking about the ecology of media means thinking about place, not only in the past, but in the present and in the future.

Concerns about book biodeterioration, which are, in essence, concerns about book survival, are also, at bottom, concerns about place. Since agreed-upon "ideal conditions" facilitate temporary loans across collections, it might seem to be the case that the place of the archival library is inconsequential.[48] Yet while conditions may be extremely similar across archives and libraries, the cost of maintaining supposedly ideal conditions will differ based on local climate.[49] It is useful, then, to think about these preservation spaces existing in biomes, which the World Wildlife Fund (WWF) describes in this way: "Biomes are the various regions of our planet that can best be distinguished by their climate, fauna and flora. There are different ways of classifying biomes but the common elements are climate, habitat, animal and plant adaptation, biodiversity and human activity."[50] The WWF designates five major biome classifications: Aquatic, Desert, Forest, Grasslands, and Tundra.[51] By the broadest, five-biome categorization standards, a book can cross the Atlantic Ocean and remain in the same biome type. For instance, to look at a copy of Shakespeare's *Sonnets* in the Folger Shakespeare Library in 2016 is to hold that volume in the same biome type, Forest (and more specifically Temperate Deciduous Forest), in which it was printed in 1609. However, move the book to the Huntington

Library, and you will be reading Shakespeare in a Desert biome, even though conducting research in these two reading rooms, Folger and Huntington, feels similar in terms of climate.

Of course, we do not find ourselves in anything like a forest when exiting the Folger. (Thomas Thorpe, printer of Shakespeare's *Sonnets*, would have been equally surprised in 1609 if someone told him his print shop was in a forest.) Biomes designate regions in which climates (and plants and animals) differ significantly, and as such, they are broad categories largely focused on nonhuman elements. In 2007, Erle C. Ellis and Navin Ramankutty called attention to the fact that the biome classification systems tend to "ignore humans altogether or simplify human influence." Their solution was a system of "terrestrial biomes based on global patterns of sustained, direct human interaction with ecosystems," which they called "anthropogenic biomes" or "anthromes."[52] Ellis and Ramankutty identify six anthromes, all of which include subgroups: Dense Settlements (also sometimes referred to as Urban), Villages, Croplands, Rangelands, Forested, Wildlands. Regardless of whether we categorize by biome or anthrome, London ca. 1609 and Washington, DC, ca. 2009 remain in the same category: a Forest biome or an Urban anthrome. While anthrome classifications allow a bit more specificity, neither anthrome nor biome really captures the sameness that can exist inside the British Library or the Folger or the Huntington. All of which is to say that giving "a local habitation and a name" to an archival library is more challenging than it seems.

While archival libraries exist within distinct biomes and anthromes, the spread of climate-controlling systems—or what the air-conditioning pioneer Willis Carrier called "man-made weather"—also allows them to exist within micro-environments or what can be best classified as microbiomes. The term "microbiome" can indicate a range of micro-environments, from the human microbiome (a topic on which quite a lot of research has been done) to the microbiomes of built/indoor environments (a topic that has attracted significant attention in recent years).[53] The National Academies of Sciences, Engineering, and Medicine explains that within built environments

> there exist a vast number and diversity of species of bacteria, viruses, fungi and protozoa in the air, water, and heating, ventilation, and air conditioning (HVAC) systems, and on surfaces. These constitute dynamic microbial communities or "microbiomes." The nature, composition, diversity, evolution, and growth of these microbiomes are influenced by interactions with humans, animals and plants, and by factors

> such as air flow, temperature, humidity, chemical exposures and building materials. These factors are, in turn, shaped by the design, construction, operation, occupation, and use of the built environments.[54]

What is intriguing is that a focus on microbiomes actually draws attention back to the broader biomes and anthromes in which a built environment exists and the ways that the microbial profile of a given indoor space is influenced by its immediate outdoor environment.[55] In other words, while temperature and relative humidity profiles might be standardized across institutions to create supposedly ideal conditions, the microbial profiles of libraries and archives will differ drastically based on a host of factors that are just now beginning to be understood by researchers. One fascinating example is research into the role of plants in built environments that suggests that "indoor plants can act as a simple but efficient way to stabilize and increase diversity of beneficial microbes in the built environment and other enclosed systems."[56]

If priority is given to preserving as many rare books as possible for the foreseeable future, then the easiest thing is also the safest: keep using the climate-control resources that have been working so far.[57] As scholars researching biodeterioration in libraries and archives have demonstrated, "If the thermohygrometric parameters [temperature and relative humidity] are set correctly, the risk of biodeterioration of materials (with the exclusion of damage caused by insects) in libraries provided with climate control systems that include air filtering is practically nonexistent."[58] The fantasy of textual fixity, in a manufactured climate where biodeterioration is "practically nonexistent," is plausible, not fantastical. Except, as the authors note, in the case of "exceptional circumstances."[59] But there is the question of what we lose in imagination when we do not lose materials. In an insightful reconsideration of Ben Jonson's "An Execration upon Vulcan," Adam Smyth reads the 216-line poem that Jonson wrote after his library was destroyed by fire in 1623 not as—or not merely as—a lament about book destruction, but as a meditation on the important role of a book's environmental vulnerability. The poem, according to Smyth, is a "plea . . . for destruction to play a trimming, quasi-editorial role in the construction of the aesthetic and bibliographical category we call 'literature'. The 'Execration' might thus be read as a meditation on the uses of destruction for the preservation of books."[60] To return to the technical and poetic questions of survival and decay with which this chapter began, we again see how the technical and poetical are intertwined. We might think of *Sonnet* 64:

> Ruin hath taught me thus to ruminate,
> That Time will come and take my love away.

This thought is as a death, which cannot choose
But weep to have that which it fears to lose. (64.11–14)

To lose the fear of ruin, to lose the fear of loss, is also to make the very real (in the Renaissance) conflict between time and decay appear cartoonish.

The staging of a battle between time and paper memorialization may be a poetic cliché, but, like a proverb, such a cliché arises from lived ecological reality. As Smyth observes, the history of Renaissance books "is also a history of loss and destruction. . . . Most printed texts lived very briefly, and then were gone, forever."[61] If pages from a 1609 first edition of Shakespeare's *Sonnets* were used for toilet paper in 1609, it might be considered an action entirely compatible with human practices at the time. If pages from a 1609 *Sonnets* were used for toilet paper in 2016, we would call it a catastrophe. We have, as best we can, halted the decay of extant Shakespearean volumes such that catastrophe is their only chance at decomposition. Cataloguing all the defects of torn, scuffed, foxed, and blotted books as if these marks are abnormal, imagining "ideal" copies of Renaissance texts as normal, and supposing that these "ideal" copies are stable objects that can be indefinitely protected from decay, we miss out on the urgency of the poetry they hold. We fool ourselves, too: there is little chance that Shakespeare's *Sonnets* will endure in this paper format for another four hundred years. The truth is more poetic.

Hueck's definition of biodeterioration is one that puts emphasis on value; biodeterioration is "any undesirable change." It is a definition that also usefully draws attention to the role of imagination and fantasy, though, for the change we do not want is caused by "the vital activities of organisms." Stefan Michalski, in his contribution to the Getty Conservation Institute forum that I discussed at the beginning of this chapter, further emphasizes the roles of value and imagination in our acts of preservation: "The uncertainty and the difficulty in these risk assessments do not reside in the material science. . . . The largest uncertainty resides in estimates of the loss of value."[62] In this closing section, titled "Toward Assessment and Acceptance of Loss," Michalski suggests that one of the great challenges of making archival preservation sustainable is "the belief in an ideal collection" and in "perfect preservation." The problem, his language suggests, is more poetic than pragmatic: it's not that perfect preservation "was ever a reality but that it was a great comfort."[63] Imagining we have a choice between preservation and decay, imagining we can rely on "combustible material . . . formed from remains of organisms during the geological past," that is, fossil fuels (and nuclear energy and water resources), to preserve first editions of Shakespeare's *Sonnets* right through the Anthropocene, we may bundle up with our books in an "ideal climate" of 21° Celsius and 50 percent relative humidity.[64]

But biodeterioration keeps books alive, wear and tear keep books alive, and the fear of loss and decay more than the certainty that we are insulated from it keeps books alive.

Inside: The Poetry of Loss

Inside the cool Folger vaults, where biodeterioration is being kept at bay en masse even on the most unbearably hot and humid day in August, a unique, single leaf of handmade paper cheats time. The page is remarkable in part because the words on it are John Donne's epitaph for his wife, and it was long believed that the handwriting was Donne's. It is remarkable, too, because if any Renaissance poet were to appreciate Hueck's definition of biodeterioration, with all its desire and science and vital activities and subject positions, it would be Donne. As I discussed in Chapter 2, Donne thought about his paper, and he worried about its impermanence. He may have even known, to some degree, that as one study of fungal spoilage in historical papers puts it, "Paper, before being the constituent of ancient books and precious manuscripts, is an organic material, and therefore a source of nourishment for many organisms."[65] Mark but this fungi.

The very existence of this copy of Donne's poetic epitaph for his wife (known as the Loseley manuscript of the "Epitaph for Ann Donne"), handwritten on an environmentally susceptible medium, is remarkable. Most remarkable, though, in the context of a conversation about textual decomposition, about stone biodeterioration and paper preservation in archival libraries, is the way this poem's paper is linked, both rhetorically and ecologically, with stone. In the first place, it is a paper epitaph, an epitaph composed to be displayed on a tomb but ultimately transmitted to the present day on paper.[66] A solitary word, "Annæ," appears at the top, just left of center and surrounded on all sides not by white space, but by a tawny colored space filled with flecks of organic matter.[67] The left diagonal of the *A* and the finial of the *e* in *æ* are subtly interrupted by these flecks, visible to the naked eye.[68] Upon closer inspection, the page is filled with hair. Little flecks here and there, probably from the felts used to couch the paper as it was being made. The presence of hair is hard to miss in a poem about Donne's dead beloved, especially when Donne's earlier poems imagine the decay of a beloved's hair in the grave.[69] "The Funeral," for example, begins, "Whoever comes to shroud me, do not harm / Nor question much / That subtle wreath of hair, which crowns my arm," and revolves around the life and afterlife of the hair bracelet.[70] "The Relic" also takes as its subject a hair bracelet in the grave:

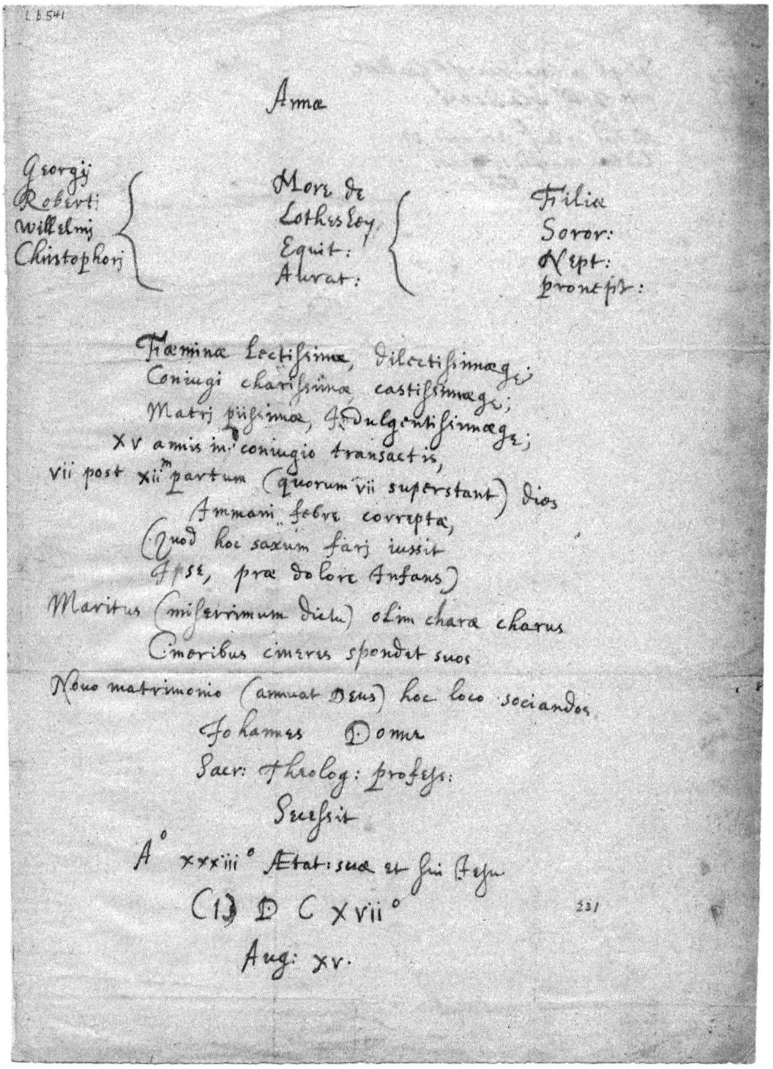
L.b.541

Annae

Georgii
Roberti
Wilhelmi
Christophori

More de
Lothesley
Equit:
Aurat:

Filiae
Soror:
Nept:
Pronept:

Foeminae Lectissimae, dilectissimaeque;
Coniugi charissimae castissimaeque;
Matri piissimae, Indulgentissimaeque;
XV annis in coniugio transactis,
vii post xii^m partum (quorum vii superstant) dies
Immani febre correpta,
(Quod hoc saxum fari iussit
Ipse, prae dolore Infans)
Maritus (miserrimum dictu) olim chara charus
Cineribus cineres spondet suos
Novo matrimonio (annuat Deus) hoc loco sociandos
Iohannes Donne
Sacr: Theolog: Profess:
Secessit
A° xxxiii° Aetat: suae et sui Jesu
CIↃ DC xvii°
Aug: xv.

FIGURE 30. John Donne's "Epitaph for Anne Donne [manuscript], 1601/1602 February 2," L.b.541, recto. Used by permission of the Folger Shakespeare Library.

When my grave is broke up again
Some second guest to entertain,
. . . .
And he that digs it, spies
A bracelet of bright hair about the bone,
                    Will he not let us alone,
And think that there a loving couple lies,
Who thought that this device might be some way

To make their souls, at the last busy day,
Meet at this grave, and make a little stay?[71]

Though I knew them to be nonhuman, the hairs in the paper had a dizzying effect on me, like the moment I first realized I was holding dead animals in archival libraries and wondered how I could have overlooked them. When I thought about the hair, when I thought about the hand that wrote the epitaph, I was overwhelmed with the sudden realization that I was reading about a tragic, traumatic experience for the surviving father and children. And that I was reading about someone who never lived to be as old as me. Two of them, in fact: mother and child. The Loseley manuscript of "Epitaph for Ann Donne" is a material treasure. In editions of Donne's poetry that include "Epitaph for Ann Donne," it is generally relied upon as the copy text. It may be one of a very few poems we have in Donne's hand.[72] But it is also a memorial of the real pain and grief experienced by Donne upon losing a child and then, seven days later, the mother, "a lady most rare, most beloved."[73]

The leaf of paper known as the Loseley manuscript is one of five known manuscript copies of an engraved epitaph that no longer exists. The stone epitaph, carved by engraver Nicholas Stone, was once "A faire Monument in the Chancell, on the North side, at the upper end" in St. Clement Danes Chapel, according to John Stow, who recorded the epitaph in print in his 1633 *Survey of London*.[74] The stone inscription "did not survive the London fire of 1666 and the rebuilding of St. Clement Danes in 1680–82."[75] The only reason we know what Donne wrote for his wife's tomb is because paper has preserved the words that were carved into stone. Milgate translates "(Quod hoc saxum farj iussit / Ipse, prae dolore Infans)" in lines 12–13 as "(at which he himself, speechless when confronted with the sorrow, commanded this stone tablet to speak)."[76] To read the paper is to mix media or to see media mixed: surviving paper mediates lost stone tablet which was meant to mediate lost speech.[77]

Two of the five other known manuscript copies of "Epitaph for Ann Donne" add this note after the epitaph: "In the chancell of St Clement Danes church without Temple Barre: made by J: D: himselfe, who was afterwards notwithstanding buried in Paules whereof he dyed Deane: A: D: i63i."[78] Unlike Ann's epitaph, we need no paper to tell us John Donne's epitaph: the words survive as Nicholas Stone first engraved them in the south quire of St. Paul's Cathedral. Below the engraving, according to Ann Rosalind Jones and Peter Stallybrass, "John Donne strikingly displays his winding sheet in what was probably then, as now, the most famous English monument of the early seventeenth century."[79] The effigy of Donne is, like the bas-reliefs of Romeo and Juliet at the Folger Shakespeare Library, done in white marble. Donne's statue has fared better than the star-crossed lovers, though; according to St. Paul's website, it is "one of the few monuments to have survived the Great Fire of London," though "scorch marks can be seen on its base."[80] John Stubbs claims it is "the only piece of sculpture to survive the Great Fire in decent condition."[81] Before the Great Fire of 1666, Stubbs explains, Donne's monument survived the political upheaval of the English Civil War, when "Parliamentary forces, garrisoned in the Cathedral . . . had exercised their horses in the aisles."[82] Three hundred years after the Civil War, it remained intact through the Blitz of World War II.

Donne's effigy lives on paper, too. Engraved and printed as a frontispiece to *Deaths Dvell*, Donne's last sermon, and, according to the title page, "The Doctors owne Fvnerall Sermon."[83] It is this stone memorial and its representation in print that John Marriot gestures toward in the first printed edition of Donne's poems (1633):

Hexastichon Bibliopolæ
I see in his last preach'd, and printed booke,
His Picture in a sheete; in *Pauls* I looke,
And see his Statue in a sheet of stone,
And sure his body in the grave hath one:
Those sheetes present him dead, these if you buy,
You have him living to Eternity.[84]

According to Marriot, then, a potential reader/buyer (especially one shopping in the bookstalls outside St. Paul's) finds Donne memorialized in four sheets:

1. Lines 1–2: the sheet of paper on which the engraved frontispiece appeared in *Death's Duel.*
2. Lines 2–3: The white marble winding sheet depicted on the statue in St. Paul's Cathedral.
3. Line 4: The cloth winding sheet that "sure his body in the grave" must be wearing.
4. Lines 5–6: The sheets of paper that Marriot (and printer Miles Fletcher) present as *Poems* (1633).

Winding sheets, like paper, were made from linen, and as Sarah Wall-Randell notes, since winding sheets "might well be taken directly off the deceased's bed . . . winding-sheets were used sheets, making them even more like raw material of paper."[85] Marriot's poem may be a unique, extant example of sixteenth- and seventeenth-century writing that directly links winding sheets and papermaking—a notable fact when we consider, as discussed in the previous chapter, how often Renaissance readers, printers, and writers registered the direct relationship between cloth and paper.[86] But if it is unique in terms of its reference to Renaissance material culture, Marriot's poem is commonplace in terms of its reference to Renaissance memorials as mixed media. Reading "Hexastichon Bibliopolæ," we find that Donne's sermon (delivered orally) and Donne's stone statue have been remediated and rememorialized with an iron pen of sorts—the printing press—which records word and image on sheets of paper. The "sheet of stone" that signifies both Donne's statue and the sheet in which Donne is wrapped in the stone sculpture is contrasted with Donne's actual body, which, wrapped in a winding sheet below ground, is actively decomposing. In place of a decomposing Donne, Marriot offers Donne newly gathered in printed sheets.

But what of the book imagined in "Of the Book"? We do not have the manuscripts to which Donne refers in "Of the Book"; the letters between Ann and John Donne have not survived, letters that, if Donne's claim is to be believed, might have been gathered into a singularly touching exemplum of love. In fact, no letter by Ann Donne survives at all, a reality that Stubbs calls "infuriating and scandalous."[87] Sitting in archival libraries in the twenty-first century, frigid fingers handling a page that Donne's seventeenth-century fingers may have held before me, I find it hard to generate the heat for infuriation. From an ecological perspective, rather than being scandalized by the loss of Ann Donne's papers, we might be surprised by the survival, despite biodeterioration, fires, and bombings, of John Donne's papers. At the end of a wonderfully informative primer to researching Donne in archival libraries, Lara M. Crowley concludes that "the wealth of surviving manuscript copies of Donne's works presents exceptional opportunities to study the nature of textual transmission. . . . As we strive to illuminate Donne's poetry and prose, we must obey his own command in ["Of the Book"]: 'Study our manuscripts' (l. 10)."[88] In doing so, we must study the "nature of textual transmission" not only in an abstracted, bibliographical sense, but also in a material, ecological sense. The nature of textual transmission now includes our active, intentional attempts to stop decay, and we have HVAC (and fire retardant) systems to thank, at least in part, for the "wealth of surviving manuscript[s]."

Donne seems not to have expected his actual love letters with Ann to survive: the afterlife that "Of the Book" imagines for the poet and his beloved is not one in which we read manuscripts of love letters, but one in which we read a more durable book object made from manuscripts, one that can "out-endure" the vagaries of time (l. 5). Ann Donne's epitaph, well preserved on paper inside a marble building whose stone monuments of love are quickly biodeteriorating, is what we have instead. It is, M. Thomas Hester claims, "one work . . . we know for certain that Donne wrote for his wife."[89] "Epitaph for Ann Donne" serves, then, as evidence and example not only of the love shared between poet and wife, but also of the way records of such love might be effectively preserved. In contradistinction to the durable object imagined in "Of the Book," "Epitaph for Ann Donne" suggests the advantages of a more duplicitous media preservation strategy.

Texts can never really be disentangled from the ecosystems in which they come to be and in which they will cease to be. When we organize books, and our ideas about books, into categories and systems that do not account for the organic nature of the page, we overlook data that can help us to better understand and preserve the Renaissance books we still have with us. To read ecosystemically is to tease out the worlds of allusion and wordplay and historical context and environmental negotiation from the plant fibers of centuries-old handmade paper. It is to begin to understand the roles of scarcity and possibility and corruptibility in the stories that we are told and in the stories that we tell about the past.

REMAINDERS

Reading and Seeing Textual Ecology

> HAMLET. Dost thou think Alexander looked o' this fashion i' th' earth?
> HORATIO. E'en so.
> HAMLET. And smelt so? Pah! [*He puts the skull down*]
> HORATIO. E'en so, my lord.
> HAMLET. To what base uses we may return, Horatio! Why may not imagination trace the noble dust of Alexander till he find it stopping a bunghole?
> HORATIO. 'Twere to consider too curiously to consider so.
> HAMLET. No, faith, not a jot; but to follow him thither, with modesty enough and likelihood to lead it, as thus: Alexander died, Alexander was buried, Alexander returneth to dust; the dust is earth; of earth we make loam; and why of that loam whereto he was converted might they not stop a beer barrel?
>
> —*Hamlet*, 5.1.204–19

Horatio may be a pragmatist, and there may be more things in heaven and earth than are dreamt of in his philosophy, but he makes a valid point in response to Hamlet's imaginative reading of Alexander the Great's dust. He and Hamlet are friends, but they are also scholars, and one must draw a line between an insightful close reading and a willfully inventive reading. Intriguingly attentive to ecology as it may be, Hamlet's fanciful etiology of Alexander's dust is contrived. Hamlet's zero-sum narrative is also inventively neat: Alexander's dusty but unscattered remains cohere across time and space so that one historical figure's dust = one beer stopper. Horatio's critique is not that the method is invalid, but that the mode is too sweeping: "'Twere to consider too curiously to consider so."

In *The Nature of the Page*, I have tried to draw on the imaginative, poetic potential of guessable backstories, but I have also tried to limit my purview in ways that allow careful, detailed re-memberings of the past lives of texts. And, rather than imagining zero-sum narratives that unify all bits and pieces into tidy textual forms, I have found a special joy in those remainders that do not quite fit into unifying narratives. Translation, whether from one language to another or from mental idea to words on a material page, is messy. It renders likeness, not exactness. Translating ideas into material texts allows them to be stored and transmitted, but it also alters those ideas and creates remainders—material evidence on and in the page in the forms of plant fibers, plant husks, human and nonhuman hair, accidental and intentional blots, gelatinized animals, river silt. These visible bits of material evidence were legible to readers of the past. Their exact origins may have been unknown, but they were for the most part guessable.

These remainders remain visible, if not legible, to readers now. As Chapter 1 demonstrated, the progress of papermaking from then to now is as troubling as it is triumphant; it relies on ecological exploitation and forced labor, on the ship sails of colonial agents and on the spinning wheels of slaves. Plant fibers embedded in old paper still offer up new accounts of these troubling environmental histories. Chapter 2 explored a historical methodology for finding the lives of plants, animals, and linen-wearing readers in pages of Renaissance books. Chapter 3 insisted on reading blots, even accidental ones, as a means of discovering how media technology changed material reading and writing habits and how, by extension, those material habits inflected everything from Shakespeare's history plays to theologically significant biblical metaphors. Finding animals hiding in plain sight in the form of animal sizing, Chapter 4 drew attention to the assumptions we have been making about readerly marginalia and about book survival. Finally, Chapter 5 drew attention to the material and metaphorical implications of the climate-controlled microbiomes where Renaissance books are preserved from decay. There I suggested that it is the near absence of change, the stalling of decay, that tells a new and remarkable story about the ecology of texts.

To suggest that these narratives in and on early handmade paper are legible is to say that they are able to be read. But relying on a deeper sense of *legible*'s etymology, I also mean to suggest that the ecological elements of texts are selectable and that meaning can be discovered in the process of selecting and gathering them together. At the end of *The Divine Comedy*, in canto 33 of *Paradiso*, Dante sees "ingathered, bound by love in one single volume, that which is dispersed in leaves throughout the universe."[1] As Roger Chartier and Peter Stallybrass explain, "Dante's word for binding together, *legato*, is derived from the Latin *legere*, which means both to gather and to read. Gathered together, the leaves that had previously been scattered become legible or readable."[2] *Legible* and *selectable* share an etymology; the acts of choosing and of understanding go hand in hand. The remainders, the things that do not fit into narratives we have told in the past about books and the environments in which they come to be (and, eventually, not to be), suggest other ways of gathering and reading. More than anything, an ecology of texts is an invitation to take up new possibilities that are as engaging and playful as they are timely and necessary.

The Nature of the Page strives to bring the acts of reading and seeing into a more holistic mode of textual analysis, one that acknowledges and explores the broader ecosystemic context in which texts exist. Henry Vaughan's poem "The Book" first called my attention to this broader context. I felt foolish for not knowing that early handmade paper was sourced not from trees but from recycled clothing which was sourced from flax and hemp plants. Looking for book history scholarship that could tell me more, I found Adrian Johns's *The Nature of the Book*. In retrospect, it was the perfect find: it taught me much I needed to know about book history without answering the ecological questions that seemed most pressing to me. The title of my book, *The Nature of the Page*, acknowledges a debt of gratitude to Johns's book for all it taught me and all it made me find on my own. The goals of both books align. Both argue that there is more to the story than our categories allow, more than is dreamt of in our histories and methodologies. In the Introduction, Johns claims that the work of printing is "dedicated to effacing its own traces"; his book "tries accordingly to excavate the complex issues involved in the historical shaping of print—issues that our conventional notion of print culture obscures with all the authority of a categorical definition."[3] My hope is that the reader who has made it this far cannot now read words such as "effacing" and "excavate" in the quotation above without thinking about the sorts of ecological negotiations that text-making and text-management requires.

If readers of this book can never again look at paper, especially in an archival library, and see it as blank or white, then *The Nature of the Page* is at least partially successful. But there is work to do that goes beyond the beguiling recognition that an old Bible's paper might be made from the clothes of an atheist and that a beer stopper might be composed of Alexander's dust. Had events unfolded differently, Hamlet the poet and Horatio the historian might have written a tremendous study—"Alexander's Afterlives: From Babylon to Bungholes," perhaps—because of, not in spite of, their differences. The work of tracing textual ecologies is, I have found, insistently collaborative, and it rewards the kinds of conversation that arise from misunderstanding and confusion. So much of what counts here as my book arises from the questions that presented themselves or facts I learned while looking with colleagues and friends and students at old books in archival libraries, while getting my hands and most of the rest of myself wet in papermaking studios, while traveling along the Danube River and the Mississippi River with an international collective of scholars affiliated with the Center for Culture, History, and Environment and with the Rachel Carson Center for Environment and Society, and while touring the facilities of those who care for and conserve paper. These kinds of intentional collaborations are necessary, for an ecology of texts demands that varied ways of seeing and varied ways of reading come into extended contact.

In "The Emergence of the Environmental Humanities," Robert S. Emmett and David E. Nye emphasize the localized sites that have shaped the development of the environmental humanities. They write that "because of its wide range, the field of environmental humanities is difficult to pin down, and it has different profiles depending on the scholarly strengths at the institutions where it has emerged."[4] Archival libraries are the institutions in which my book has emerged because they are dynamic, fecund sites for meaningful, innovative collaborations that can help shape—and be shaped by—book history and environmental humanities conversations that are devoted to more deeply reading and seeing the records that, at least for now, remain extant. They are unique spaces filled with unique data about the histories of ideas and about the animals, vegetables, and minerals on which those ideas have been recorded. However they may come to be reimagined and redesigned in response to a changing climate, archival libraries seem to me to be the base camps from which crucial environmental humanities and book history collaborations will depart and return.

For all my best attempts to more fully comprehend textual ecology, *The Nature of the Page* is a case study. I had to narrow my focus to paper because there was just too much to discover about textual ecology in Renaissance England alone, not to mention what is left to learn about the textual ecologies of other periods and places. Minerals, for example, which were integral to Renaissance reading and writing practices, are largely absent from this study, though they were also essential components of the computers, tablets, and smartphones that made these words on this page possible. Once a point of writerly anxiety, my inability to write comprehensively about textual ecology now inspires scholarly anticipation. The ecopoetic story repeats, everywhere and always: filled with inspiration and insight, a poet—or a mathematician, a father, a daughter, a prisoner, a refugee, a web developer, a street artist, a grandmother, a teenager, a microbiologist, a songwriter—reaches for something near at hand, some assemblage of nonhuman materials, that might allow human ideas to be held in their hands and shared with others. Singled out from the many ecological narratives that might be told about clay tablets, birch bark maps, Victorian headstones, or email spam, *The Nature of the Page* suggests some of the many ways we might make the natural fabric of their stories and our stories more legible.

NOTES

PREFACE

Note to epigraph: William Shakespeare, *As You Like It*, 2.1.16–17, in *Shakespeare's Plays, Sonnets, and Poems* from Folger Digital Texts, ed. Barbara Mowat, Paul Werstine, Michael Poston, and Rebecca Niles (Washington, DC: Folger Shakespeare Library, n.d.), www.folgerdigitaltexts.org. Except where a specific edition is indicated, all Shakespeare quotations in this book are cited from this edition.

1. Shakespeare, *Sonnets*, 12.13.

2. Claims about toxins are based on Apple's 2017 Environmental Responsibility Report (covering the fiscal year 2016); see "Eliminating Toxins," in Apple Inc., *Apple Environmental Responsibility Report 2017*, April 2017, http://images.apple.com/environment/pdf/Apple_Environmental_Responsibility_Report_2017.pdf, 31. On the undisclosed materials used in making iPads and Kindles, see Daniel Goleman and Gregory Norris, "How Green Is My iPad?" *New York Times*, April 4, 2010. See also Erika Engelhaupt, "Would You Like That Book in Paper or Plastic?," *Environmental Science and Technology* 42, no. 12 (June 15, 2008): 4245.

3. Recovery figures for finite resources are from Apple's 2016 Environmental Responsibility Report. The precise figures are 189,544 pounds of cobalt, 2,204 pounds of gold, and 4,518,200 pounds of aluminum; see Apple Inc., *Apple Environmental Responsibility Report 2016*, September 2016, https://images.apple.com/environment/pdf/Apple_Environmental_Responsibility_Report_2016.pdf, 17. The 2017 report also indicates that "for every 100,000 iPhone 6 devices" Apple is able to recover 24kg of "Rare Earth Elements"; see Apple Inc., *Apple Environmental Responsibility Report 2017*, 18.

4. "The Adirondack Park was created in 1892 by the State of New York amid concerns for the water and timber resources of the region." See "About: The Adirondack Park," Adirondack Park Agency, Apa.ny.gov, http://apa.ny.gov/About_Park/index.html. On New York City as the beneficiary of the "diversified local hinterland, with expanding productions from farms, forests, and iron mines upriver," see Donald William Meinig, *The Shaping of America: A Geographical Perspective on 500 Years of History*, vol. 1 (New Haven: Yale University Press, 1986), 126, 119–29.

5. Aldo Leopold, *A Sand County Almanac*, Leopold Pines Edition (Baraboo, WI: Land Ethic Press, 2007), 24–25. See also Aldo Leopold, *A Sand Country Almanac* (New York: Oxford University Press, 1949), 24–25. Throughout this work, I read nature writers as, and alongside, theorists. All of these writers, from Leopold to Latour, can and do at times settle for totalizing narratives or unfounded claims; all of them, too, offer modes of inquiry that productively unsettle anthropocentric narratives of the past by emphasizing negotiations among actors or biota or things.

6. Leopold, *Sand Country Almanac* (2007), 205, 25.

7. Leopold, 25.

8. Leopold, 25.

9. Leopold, unnumbered flyleaf.

10. Leopold, unnumbered flyleaf.

INTRODUCTION

Note to epigraph: Hannes Bergthaller, Rob Emmett, Adeline Johns-Putra, Agnes Kneitz, Susanna Lidström, Shane McCorristine, Isabel Pérez Ramos, Dana Phillips, Kate Rigby, and Libby Robin, "Mapping Common Ground: Ecocriticism, Environmental History, and the Environmental Humanities," *Environmental Humanities* 5, no. 1 (2014): 272.

1. Margreta de Grazia and Peter Stallybrass, "The Materiality of the Shakespearean Text," *Shakespeare Quarterly* 44, no. 3 (Fall 1993): 280.

2. According to the University of Pennsylvania Press production manager, the paper is "30% certified post-consumer waste (i.e., industrial sources of waste paper), a portion of certified mixed recycled sources, and FSC [Forest Stewardship Council] chain-of-custody-certified new pulp." Jerome Singerman, personal communication to the author, May 15, 2018.

3. Thomas Dekker and George Wilkins, *Iests to make you Merie: With The Coniuring vp of Cock VVatt, (the walking Spirit of Newgate) To tell Tales*, STC 6541 (London: By N[icholas] O[kes] for Nathaniell Butter, dwelling neere to St. Austins Gate, at the signe of the pide Bull, 1607), 61; H3r.

4. William Shakespeare, *Sonnets*, 108.1.

5. Sir Philip Sidney, *Sir Philip Sidney's an Apology for Poetry, And, Astrophil and Stella: Texts and Contexts*, ed. Peter C. Herman (Glen Allen, VA: College Publishing, 2001), 1.5–13.

6. See also Smyth's comparison of Strode and Sidney, which calls attention to the fact that while poets may not agree on the relationship between material and metaphor, they must grapple with it. Discussing "On Fayrford Windows," Smyth writes, "Strode's poem presents itself, then, as a poor surrogate for the reality of the windows—there is no hint of Sidney's notion of poets improving on nature, here—and, like many of Strode's poems, this is a verse that laments the limitations of verse, drawing attention to how much cannot be conveyed." Adam Smyth, " 'Art Reflexive': The Poetry, Sermons, and Drama of William Strode (1601?–1645)," *Studies in Philology* 103, no. 4 (Fall 2006): 444, 447.

7. See Robert Darnton, "What Is the History of Books?," *Daedalus* 111, no. 3 (Summer 1982): 67, 66. For a critique of Darnton's "disciplinary and historical divisions" that advocates other kinds of unifying principles such as "the acknowledgement that the materiality of a text signifies and . . . insists upon a set of audiences or responses," see Graham Allen, Carrie Griffin, and Mary O'Connell, introduction to *Readings on Audience and Textual Materiality*, ed. Graham Allen, Carrie Griffin, and Mary O'Connell (London: Routledge, 2015), 4.

8. Bergthaller et al., "Mapping Common Ground," 264; and Ursula K. Heise, "Introduction: Planet, Species, Justice—and the Stories We Tell About Them," in *The Routledge Companion to the Environmental Humanities*, ed. Ursula K. Heise, Jon Christensen, and Michelle Niemann (London: Routledge, 2017), 1.

9. Deborah Bird Rose, Thom van Dooren, Matthew Chrulew, Stuart Cooke, Matthew Kearnes, and Emily O'Gorman, "Thinking Through the Environment, Unsettling the Humanities," *Environmental Humanities* 1, no. 1 (2012): 1.

10. See Michel Serres on our tendency to forget "the world of things themselves." Michael Serres, *The Natural Contract*, trans. Elizabeth MacArthur and William Paulson (Ann Arbor: University of Michigan Press, 1995), 2.

11. Bergthaller et al., "Mapping Common Ground," 261.

12. Bergthaller et al., 272.

13. The word "utopia" was More's invention; he "coined it by fusing the Greek adverb *ou*— 'not'— with the noun *topos*—'place'—and giving the resulting compound a Latin ending." Thomas More, *Utopia*, ed. George M. Logan, and Robert M. Adams, trans. Robert M. Adams (Cambridge: Cambridge University Press, 1989), xi.

14. More, 79.

15. Thomas More, *Memoirs of Sir Thomas More, &c.*, ed. Arthur Cayley, the Younger, Esq., vol. 2 (London: Cadell and Davis, 1808), 101.

16. More, *Utopia*, 79.

17. Darnton himself acknowledges that his Communications Circuit was meant to be provisional, not authoritative. For his own critique of the circuit and his discussion of others' critiques, see Robert Darnton, " 'What Is the History of Books?' Revisited," *Modern Intellectual History* 4, no. 3 (2007): 495–508.

In an in-progress article, I discuss the circuit and surprising, seemingly accidental changes made to it in the years after it was first published.

18. Leslie Howsam, "The Study of Book History," in *The Cambridge Companion to the History of the Book*, ed. Leslie Howsam (Cambridge: Cambridge University Press, 2014), 8.

19. D. F. McKenzie, *Bibliography and the Sociology of Texts* (Cambridge: Cambridge University Press, 1999). McKenzie first delivered these ideas as Panizzi lectures in 1985 and first published them in 1986.

20. McKenzie, 29.

21. Mark Vareschi, *Everywhere and Nowhere: Anonymity and Mediation in Eighteenth-Century Britain* (Minneapolis: University of Minnesota Press, 2018), 2.

22. Elsewhere, I have referred to this as a "social ecology of texts," a term that has the advantage of responding more obviously to "sociology of texts" but that has the disadvantage of redundancy. In a literal sense, "ecology" as a human practice makes sense of (*logos*) or creates order in and from an *oikos*, or dwelling place. In that sense, "social ecology" is redundant, for ecology is inherently social. See Joshua Calhoun, "The Word Made Flax: Cheap Bibles, Textual Corruption, and the Poetics of Paper," *PMLA* 126, no. 2 (March 2011): 339–40.

23. Lucien Febvre and Henri-Jean Martin, *The Coming of the Book: The Impact of Printing, 1450–1800*, ed. Geoffrey Nowell-Smith and David Wooton, trans. David Gerard (London: Verso, 1997). The book was first published as *L'apparition du livre* (1958) and first appeared in English in 1976.

24. Darnton, "What Is the History of Books?," 66. For an excellent, concise history of the intersecting and overlapping fields of bibliography, history of the book, and the Annales school, see Martin Antonetti, "Exploring the Archaeology of the Book in the Liberal Arts Curriculum," in *Teaching Bibliography, Textual Criticism, and Book History*, ed. Ann R. Hawkins (London: Pickering & Chatto, 2006), 19–24.

25. Febvre and Martin, *Coming of the Book*, 33, 37.

26. Darnton, "What Is the History of Books?," 81.

27. Rob Nixon, *Slow Violence and the Environmentalism of the Poor* (Cambridge, MA: Harvard University Press, 2011), 96. The opening pages of Nixon's book also describe the World Bank's "scheme to export rich nation garbage, toxic waste, and heavily polluting industries to Africa," a plan that "would help ease the growing pressure from rich-nation environmentalists who were campaigning against garbage dumps and industrial effluent" (1–2).

28. Nixon, 93–96. See McKenzie, *Bibliography*, 77ff.

29. See Diane Kelsey McColley, *Poetry and Ecology in the Age of Milton and Marvell* (Burlington, VT: Ashgate, 2007), especially 113–15, where she discusses Vaughan's "The Book."

30. Archives and libraries are "distinct kinds of institutions with very different origins and natures," as Alan Galey observes in *The Shakespearean Archive*. As indicated by the title, Galey opts for "archive" in order to highlight the "irreducible duality" of the word as physical and imaginative site. However, because of "the critical tendency to describe any collection of vaguely interesting objects as an archive" that Galey observes, and because this work focuses on material interactions with textual ecologies in particular geographical spaces, I take a different tack. I use "archival libraries" in this book to indicate a primary interest in material (nonmetaphorical) archives—and in an attempt to gesture toward the irreducible duality for which Galey so convincingly argues. Alan Galey, *The Shakespearean Archive: Experiments in New Media from the Renaissance to Postmodernity* (Cambridge: Cambridge University Press, 2014), 42, 13–14.

31. Bruce Holsinger, "Of Pigs and Parchment: Medieval Studies and the Coming of the Animal," *PMLA* 124, no. 2 (March 2009): 619. See also Jennifer Borland, "Unruly Reading: The Consuming Role of Touch in the Experience of a Medieval Manuscript," in *Scraped, Stroked, and Bound: Materially Engaged Readings of Medieval Manuscripts*, ed. Jonathan Wilcox (Turnhout: Brepols, 2013), 97–114.

32. N. Katherine Hayles's *Writing Machines* (and its complementary "Lexical Linkmap" website) coins the term "Medial Ecology" as a figurative "phrase [which] suggests that the relationships between different media are as diverse and complex as those between different organisms coexisting within the same ecotome" (5). See N. Katherine Hayles, *Writing Machines* (Cambridge, MA: MIT Press, 2002). See the Lexical Linkmap at https://mitpress.mit.edu/sites/default/files/titles/content/mediawork/titles/writing/sup/sup_index.html. N. Katherine Hayles and MIT Press, "Writing Machines Web Supplement,"

MIT.edu, https://mitpress.mit.edu/sites/default/files/titles/content/mediawork/titles/writing/sup/sup_index.html.

33. For accounts of paper made from items as odd as cucumbers, see Chapter 1.

34. Bacon observes that, as a rule, artificial materials are made by weaving threads into flexible objects or by drying liquids into hard objects, but paper is something of a hybrid: "not fragile like glass nor woven like cloth; it certainly has fibres, but not distinct threads." Francis Bacon, *The New Organon*, ed. Lisa Jardine, trans. Michael Silverthorne (Cambridge: Cambridge University Press, 2000), 151–52.

35. Here I use the term "academic" to suggest not only expert knowledge published in academic journals and by academic presses, but also the sorts of institutions in which and for which such work is produced.

36. John Bidwell, "French Paper in English Books," in *The Cambridge History of the Book in Britain*, vol. 4, *1557–1695*, ed. John Barnard and D. F. McKenzie (Cambridge: Cambridge University Press, 2002), 583–601; Helen Smith, "'A unique instance of art': The Proliferating Surfaces of Early Modern Paper," *Journal of the Northern Renaissance* 8 (2017): par. 13–25, http://www.northernrenaissance.org/?p = 4847.

37. Among these, see especially Lotte Hellinga, "Printing," in *The Cambridge History of the Book in Britain*, vol. 3, *1400–1557*, ed. Lotte Hellinga and J. B. Trapp (Cambridge: Cambridge University Press, 1999), 91–97; and Andrew Pettegree, *The Book in the Renaissance* (New Haven: Yale University Press, 2011), 17–19. Mark Bland's *A Guide to Early Printed Books and Manuscripts* (Hoboken, NJ: Wiley, 2010) includes a chapter on "Paper and Related Materials," and Joseph Dane's *What Is a Book? The Study of Early Printed Books* (Notre Dame: University of Notre Dame Press, 2012) includes a chapter on "Materials: Ink, Paper." For a particularly good introductory text, one that offers an accurate, concise overview to those whose primary aim is to learn how to better see, understand, and describe early handmade paper, see Sarah Werner, *Studying Early Printed Books, 1450–1800: A Practical Guide* (Hoboken, NJ: Wiley-Blackwell, 2019), 26–34.

38. See D. C. Coleman, *The British Paper Industry, 1495–1860: A Study in Industrial Growth* (Oxford: Clarendon Press, 1958); and Richard L. Hills, *Papermaking in Britain, 1488–1988: A Short History* (London: Athlone Press, 1988). For especially good examples of scholarship that add nuance to histories of paper production and circulation, fill in gaps in the historical record, and shed new light on the many social uses of paper in early modern England, see especially Anna Reynolds, "'Such dispersive scattredness': Early Modern Encounters with Binding Waste," par. 1–43; and Smith, "'A unique instance of art,'" in "Scrutinizing Surfaces," special issue, *Journal of the Northern Renaissance* 8 (Spring 2017), https://www.northernrenaissance.org/issues/issue-8-2017.

39. Heather Wolfe, "Was Early Modern Writing Paper Expensive?," *The Collation*, February 13, 2018, https://collation.folger.edu/2018/02/writing-paper-expensive. Wolfe also delivered a talk on the topic at the Renaissance Society of America Conference, and I am grateful to her for sharing a transcript of that talk. Heather Wolfe, "How Expensive Was Writing Paper in Early Modern England?" (presentation, Renaissance Society of America Conference, Washington, DC, March 22–24, 2012). In a separate article, Wolfe, along with Roger Chartier, J. Franklin Mowery, and Peter Stallybrass, prompts conversations about paper and note-taking to include erasable surfaces such as Hamlet's "tables." See Peter Stallybrass, Roger Chartier, J. Franklin Mowery, and Heather Wolfe, "Hamlet's Tables and the Technologies of Writing in Renaissance England," *Shakespeare Quarterly* 55, no. 4 (2004): 379–419.

40. See Jonathan M. Bloom, *Paper Before Print: The History and Impact of Paper in the Islamic World* (New Haven: Yale University Press, 2001).

41. As later chapters will show, the work of media historians and those researching and theorizing the role of paper in society more broadly has also helped to shape this book. Chief among these are Ann M. Blair, *Too Much to Know: Managing Scholarly Information Before the Modern Age* (New Haven: Yale University Press, 2010) and Lisa Gitelman, *Paper Knowledge: Toward a Media History of Documents* (Durham: Duke University Press, 2014).

42. Margaret Holben Ellis, ed., *Historical Perspectives in the Conservation of Works of Art on Paper* (Los Angeles: Getty Publications, 2015), xiii.

43. Krill was senior paper conservator at Winterthur Museum for over twenty-five years. Although focused on artists' paper, Krill's research is equally useful as a guide to understanding paper used for other purposes, including printing and writing, as well as for understanding the overlapping uses of early handmade papers. See John Krill, *English Artists' Paper: Renaissance to Regency*, 2nd ed., expanded and corrected (New Castle, DE: Oak Knoll Press and Winterthur Museum, 2002), vii.

44. Sylvia Rodgers Albro, *Fabriano: City of Medieval and Renaissance Papermaking* (New Castle, DE: Oak Knoll Press, 2016).

45. For magnified fibers, see Krill, *English Artists' Paper*, 3, 7, and 134–36. It was from Krill's images that I learned to identify "shives" in paper, a topic I discuss in detail in Chapter 2. Albro's volume is filled with hundreds of high-quality, color images that repeatedly impress upon a reader the fibrous, heterogenous texture of early handmade paper.

46. See Flavio Marzi, "Fragments in Bindings," *Collection Care* (British Library blog), May 5, 2016, http://blogs.bl.uk/collectioncare/2016/05/fragments-in-bindings.html; and Iwona Jurkiewicz, "Magic in Conservation—Using Leaf-Casting on Paper and Palm Leaves," *Collection Care* (British Library blog), October 10, 2017, http://blogs.bl.uk/collectioncare/2017/10/magic-in-conservation-using-leaf-casting-a-mechanical-pulp-repair-technique-on-paper-and-palm-leaves-as-the-library-i.html.

47. See Walter Hamady, *Papermaking by Hand: A Book of Suspicions* (Mount Horeb, WI: Perishable Press, 1982), 1. *Papermaking by Hand* was hand-printed as a limited run of two hundred copies. For a more widely available, but slightly different version of Hamady's introductory thoughts on papermaking, see Walter Hamady, "Notes Regarding the Activity of Papermaking by Hand," in *Papermaking*, by Jules Heller (New York: Watson-Guptill Publications, 1978), 37.

48. Timothy Barrett, "Aesthetics and the Future of the Craft," *Hand Papermaking* 11, no. 2 (1996): 13–17.

49. Timothy Barrett et al., "European Papermaking Techniques, 1300–1800," Paper Through Time: Nondestructive Analysis of 14th- Through 19th-Century Papers, The University of Iowa, http://paper.lib.uiowa.edu/index.php. Earlier, I suggested that John Bidwell's "French Paper in English Books" was one of the best introductions to Renaissance paper; Barrett's "European Papermaking Techniques" is similarly indispensable.

50. Therese Weber, *The Language of Paper: A History of 2000 Years* (Bangkok, Thailand: Orchid Press, 2007). See also Josep Asunción, *The Complete Book of Papermaking* (New York: Lark Books, 2003).

51. The Latin translation offered here is from Bradin Cormack and Carla Mazzio, *Book Use, Book Theory, 1500–1700* (Chicago: University of Chicago Library, 2005), 1. See also William H. Sherman, *Used Books: Marking Readers in Renaissance England* (Philadelphia: University of Pennsylvania Press, 2008), xiii.

52. However, in their introduction to a special issue of *Huntington Library Quarterly*, Jennifer Richards and Fred Schurink productively question "the predilection of recent scholars for the term 'book use,'" calling for "renew[ed] attention to the 'textuality' of reading." See Jennifer Richards and Fred Schurink, "Introduction: The Textuality and Materiality of Reading in Early Modern England," *Huntington Library Quarterly* 73, no. 3 (September 2010): 345, 348.

53. Jeffrey Todd Knight, "'Furnished' for Action: Renaissance Books as Furniture," *Book History* 12 (2009): 38.

54. Smith, "'A unique instance of art'": par. 2, 3.

55. See Alyssa Connell and Marissa Nicosia, Cooking in the Archives: Updating Early Modern Recipes (1600–1800) in a Modern Kitchen, https://rarecooking.com. The Early Modern Recipes Online Collective is a collective of international scholars; the project's steering committee is Rebecca Laroche, Elaine Leong, Hillary M. Nunn, Jennifer Munroe, Lisa Smith, and Amy L. Tigner. For an example of paper research that arises from recipe research, see Marissa Nicosia's discussion of "sleek'd paper" in "To make Lemmon cakes," Cooking in the Archives, March 26, 2015, https://rarecooking.com/2015/03/26/to-make-lemmon-cakes.

56. Heidi Brayman, *Reading Material in Early Modern England: Print, Gender, and Literacy* (Cambridge: Cambridge University Press, 2005), 63.

57. Jean E. Feerick and Vin Nardizzi, eds., *The Indistinct Human in Renaissance Literature* (New York: Palgrave Macmillan, 2012), 2–3. Hooker's *Polity* (the preface and books I–IV) was first printed in 1593.

CHAPTER 1

Notes to epigraphs: Matthias Koops, *Historical Account of the Substances Which Have Been Used to Describe Events, and to Convey Ideas, from the Earliest Date, to the Invention of Paper* (London: T. Burton,

1800), 7; Margreta de Grazia and Peter Stallybrass, "The Materiality of the Shakespearean Text," *Shakespeare Quarterly* 44, no. 3 (Fall 1993): 281.

1. Gast's 1872 painting was lithographed by George Crofutt in 1873 and distributed widely. Amy S. Greenberg writes that it "was featured in his [Crofutt's] magazine, *Crofutt's Western World*" and that he "was fond enough of the image to use it as the frontispiece in his guide for western travelers." Amy S. Greenberg, *Manifest Manhood and the Antebellum American Empire* (Cambridge: Cambridge University Press, 2005), 1n2. See also George A. Crofutt, *Crofutt's New Overland Tourist and Pacific Coast Guide, Containing a Condensed and Authentic Description of over One Thousand Two Hundred Cities, Towns, Villages, Stations, Government Fort and Camps, Mountains, Lakes, Rivers, Sulphur, Soda and Hot Springs, Scenery, Watering Places, and Summer Resorts* [. . .]*: Over the Union, Central and Southern Pacific Railroads, Their Branches and Connections, by Rail, Water and Stage* [. . .] (Chicago: Overland, 1878–79), 1:300. Full text is available at https://hdl.handle.net/2027/nyp.33433081825006?urlappend=%3Bseq=285.

2. Jackson Lears calls attention to the ways in which the image has "etherealized" not only nature, but also the central female figure, and claims that the painting "accurately reflected a century-long process: the denaturing and disembodiment of female icons of abundance." T. J. Jackson Lears, *Fables of Abundance: A Cultural History of Advertising in America* (New York: Basic Books, 1995), 109.

3. To name just a few recent studies that incorporate a reading of Gast's painting into their narrative: Greenberg, *Manifest Manhood;* James David Drake, *The Nation's Nature: How Continental Presumptions Gave Rise to the United States of America* (Charlottesville: University of Virginia Press, 2011); James Joseph Buss, *Winning the West with Words: Language and Conquest in the Lower Great Lakes* (Norman: University of Oklahoma Press, 2013); Janette-Susan Bailey, *Dust Bowl: Depression America to World War Two Australia* (New York: Palgrave Macmillan, 2016); and Marisa Elena Duarte, *Network Sovereignty: Building the Internet across Indian Country* (Seattle: University of Washington Press, 2017).

4. See Dard Hunter, *Papermaking: The History and Technique of an Ancient Craft* (New York: Dover, 1947), an excellent and essential contribution to both the history of papermaking and renewed artisanal and scholarly interest in papermaking. Relevant to the concerns of papermaking vis-à-vis Western expansionism in the United States, see also Dard Hunter, *Papermaking in Pioneer America* (Philadelphia: University of Pennsylvania Press, 1952).

5. Hunter, *Papermaking*, 556.

6. On the spread of religious ideas via paper media, and on the spread of papermaking in general, see especially Jonathan M. Bloom, *Paper Before Print: The History and Impact of Paper in the Islamic World* (New Haven: Yale University Press, 2001). As Bloom notes, when we speak of a "ream" of paper, our language preserves paper's history: "The English word 'ream' comes via the Old French *rayme* from the Spanish *resma*, which itself comes from the Arabic *rizma*, meaning 'bale or bundle'" (9). The figure of "whole reams," used not infrequently in polemical rhetoric (from pamphlets and sermons to poetry and plays), subtly communicates more than even its users imagine about the material codependency of religious ideas.

7. In addition to the scholarly sources I cite throughout this book, a number of relevant trade publications on the history of papermaking have appeared in recent years—many of them successful enough to be reprinted. See Nicholas A. Basbanes, *On Paper: The Everything of Its Two-Thousand-Year History* (New York: Knopf, 2013) and its reprint (New York: Vintage, 2014); Ian Sansom, *Paper: An Elegy* (New York: William Morrow, 2013; repr., 2015); Lothar Müller, *White Magic: The Age of Paper* (Cambridge: Polity, 2015); Alexander Monro, *The Paper Trail: An Unexpected History of a Revolutionary Invention* (New York: Knopf, 2016) and its reprint (New York: Vintage, 2017); Mark Kurlansky, *Paper: Paging Through History* (New York: W. W. Norton, 2016; repr., 2017).

8. John Taylor, *The Praise of Hemp-seed*, STC 23788 (London: For H. Gosson, 1620), title page.

9. Henry David Thoreau, *Walden*, ed. Jeffrey S. Cramer (New Haven: Yale University Press, 2006), 98–99, 124, 128.

10. Thoreau, 130.

11. Jonathan Lamb describes it-narratives as "autobiographies of inanimate things" that are "episodic, usually recording a series of unhappy encounters between things and humans." Jonathan Lamb, *The Things Things Say* (Princeton: Princeton University Press, 2011), xvi.

12. Anonymous, "Adventures of a Quire of Paper," *London Magazine; or Gentlemen's Monthly Intelligencer*, August 1779, 356. On it-narratives, especially "The Adventures of a Quire of Paper," see Leah Price, *How to Do Things with Books in Victorian Britain* (Princeton: Princeton University Press, 2012);

Christina Lupton, "Gender and Materiality on the Eighteenth-Century Page," *SEL Studies in English Literature 1500–1900* 54, no. 3 (Summer 2014): 605–24; and Lupton, "The Knowing Book: Authors, It-Narratives, and Objectification in the Eighteenth Century," *NOVEL* 39, no. 3 (Summer 2006): 402–20; and Mark Blackwell, *The Secret Life of Things: Animals, Objects, and It-Narratives in Eighteenth-Century England* (Lewisburg: Bucknell University Press, 2007).

13. Anonymous, "Adventures of a Quire of Paper," *London Magazine; or Gentlemen's Monthly Intelligencer*, October 1779, 451.

14. Thoreau, *Walden*, 128.

15. Lupton, "Knowing Book," 403.

16. Thoreau, *Walden*, 129.

17. Thoreau, 133.

18. Taylor, *Hemp-seed*, 1; B3r.

19. Taylor, 19; D4r.

20. See, for example, Susan Strasser, *Waste and Want: A Social History of Trash* (New York: Metropolitan Books, 2000); and Jonathan Senchyne, "Paper Nationalism: Material Textuality and Communal Affiliation in Early America," *Book History* 19 (2016): 66–85.

21. Lyman Horace Weeks, *A History of Paper-Manufacturing in the United States, 1690–1916* (Long Island: Lockwood Trade Journal, 1916), 211.

22. Strasser, *Waste and Want*, 82. Strasser discusses the many ways early Americans, especially women, were "propagandized" into giving up their rags to the paper mills (81).

23. Weeks, *History of Paper-Manufacturing*, 59.

24. Weeks, 59.

25. Weeks, 212–18. Here Weeks cites and at times repeats lists from Joel Munsell, *A Chronology of Papermaking* (Albany: J. Munsell, 1856), v, and Charles Thomas Davis, *The Manufacture of Paper: Being a Description of the Various Processes for the Fabrication, Coloring, and Finishing of Every Kind of Paper* (Philadelphia: Henry Carey Baird, 1886), 64.

26. Philip Freneau, "On the Emigration to America and Peopling the Western Country" (1785), in *The Norton Anthology of American Literature*, ed. Nina Baym, Ronald Gottesman, Laurence B. Holland, David Kalstone, Francis Murphy, Hershel Parker, William H. Pritchard, and Patricia B. Wallace, 3rd ed. (New York: W. W. Norton, 1989), 1:719, line 4. See also John Dryden, "Palamon and Arcite," in *Fables Ancient and Modern*, Wing D2278 (London: Printed for Jacob Tonson, 1700), 1–90; B1r–N1v.

27. David McKitterick, *Print, Manuscript, and the Search for Order, 1450–1830* (Cambridge: Cambridge University Press, 2003), 183.

28. See front matter of Hunter, *Papermaking*, unnumbered flyleaf. Hunter attributes the poem to an unknown author "*circa* eighteenth century." For more on the provenance of this quotation and on the "social and metaphorical work" of rags in nineteenth-century literature, see Jonathan Senchyne, "Rags Make Paper, Paper Makes Money: Material Texts and Metaphors of Capital," *Technology and Culture* 58, no. 2 (April 2017): 545–55.

29. Robert Possehl, interview by the author, February 5, 2016. I am indebted to Possehl for sharing his unparalleled knowledge about plant harvesting and papermaking, for allowing me to participate in the processes of papermaking from plants, and for answering, over the course of four years, hundreds of technical questions about hand papermaking.

30. See Walter Hamady, *Papermaking by Hand: A Book of Suspicions* (Mount Horeb, WI: Perishable Press, 1982), 1.

31. For a useful distinction between rags as a secondary raw material and plants as a primary raw material, see Therese Weber, *The Language of Paper: A History of 2000 Years* (Bangkok, Thailand: Orchid Press, 2007), 194n69.

32. *OED Online*, s.v. "cellulose, *n.*," December 2015, Oxford University Press, http://www.oed.com.

33. Because "cellulose" is a nineteenth-century word, early hand papermakers would have also referred to plants or plant fibers, but they would have similarly understood that, for instance, paper made from flax is not made from the whole flax plant but from a particular part of the flax plant.

34. The word "cellulose" does not appear in most of the essential book history resources I use for my own scholarship and teaching and recommend to students; these resources include Joseph A. Dane, *What Is a Book? The Study of Early Printed Books* (Notre Dame: University of Notre Dame Press, 2012); D. C. Greetham, *Textual Scholarship: An Introduction* (New York: Garland, 1994); Leslie Howsam, *Old*

Books and New Histories: An Orientation to Studies in Book and Print Culture (Toronto: University of Toronto Press, 2006); R. B. McKerrow, *An Introduction to Bibliography for Literary Students* (Oxford: Clarendon Press, 1928); Andrew Pettegree, *The Book in the Renaissance* (New Haven: Yale University Press, 2010); Michael F. Suarez and H. R. Woodhuysen, *The Book: A Global History* (Oxford: Oxford University Press, 2013); and G. Thomas Tanselle, *Bibliographical Analysis: A Historical Introduction* (Cambridge: Cambridge University Press, 2009).

35. Mark Bland, *A Guide to Early Printed Books and Manuscripts* (Hoboken, NJ: Wiley, 2010), 24; John Carter and Nicolas Barker, *ABC for Book Collectors*, 8th ed. (Newcastle, DE: Oak Knoll Press, 2004), 78.

36. Josep Asunción, *The Complete Book of Papermaking* (New York: Lark Books, 2003), 20. Of cellulose, Asunción writes that it "is like the skeleton of plant cells" (103).

37. See Bloom, preface to Therese Weber's *The Language of Paper*, xv; and Weber, *The Language of Paper*, 22. See also Bloom's full volume, *Paper Before Print*, 2–3, where Bloom includes a succinct overview and summary of cellulose and its role in papermaking.

38. Weber, *The Language of Paper*, 67–68.

39. Richard L. Hills, *Papermaking in Britain, 1488–1988: A Short History* (London: The Athlone Press, 1988), 218. This volume, like Bloom's, brings together the experience of an artisan with the approach of a historian. Despite its regional focus, it is valuable as a general resource that recently gained new life as a Bloomsbury Publishing reprint in 2015.

40. The Rare Book School has long recognized the value of joining academic and artisanal interests in the study of papermaking, as attested by the course "The History of European and American Papermaking," cotaught by Timothy Barrett and John Bidwell since 1987. Sujata Iyengar's "Intermediating the Book Beautiful: Shakespeare at the Doves Press," which was inspired in part by her hands-on study of the book arts and printmaking, is a recent, exemplary case for such collaboration. It is unsurprising that academics writing about historical handmade material texts have much to learn from artisans who make material texts by hand using historical techniques. So while I am not suggesting a new kind of collaborative methodology, I am suggesting that such collaborations have a special and unrecognized potential to put book history and the environmental humanities into collaborative and sometimes even public workshops that are rigorous and intellectually generative, but that also create spaces for the sort of broad, cross-community conversations that are so vital to the work of environmental humanities scholars. See Sujata Iyengar, "Intermediating the Book Beautiful: Shakespeare at the Doves Press," *Shakespeare Quarterly* 67, no. 4 (2016): 481–502.

41. Timothy Barrett, "European Papermaking Techniques 1300–1800," Paper Through Time: Nondestructive Analysis of 14th- through 19th-Century Papers, The University of Iowa, last modified July 31, 2018, http://paper.lib.uiowa.edu/european.php.

42. Barrett, "European Papermaking Techniques."

43. Hamady, *Papermaking by Hand*, "Provisory Pre/face" (unnumbered).

44. John Gerard, *The Herball, or, Generall Historie of Plantes*, STC 11752 (London: Printed by Adam Islip, Ioice Norton, and Richard Whitakers, 1636), 949–50; Kkkk5r.

45. Bloom, *Paper Before Print*, 3. On cellulose content and fiber types, see also Robert R. A. Higham, *A Handbook of Papermaking* (London: Oxford University Press, 1963), 8–18. Higham puts the cellulose content of cotton at 92 percent, flax at 80 percent, and hemp at 77 percent.

46. John Krill, *English Artists' Paper: Renaissance to Regency*, 2nd ed. (New Castle, DE: Oak Knoll Press and Winterthur Museum, 2002), 3.

47. Koops, *Substances* (1800), title page.

48. The second edition of *Substances*, published in 1801, expands the 1800 edition from 91 pages to 273 pages. The 1801 edition prints fewer words per page, which accounts for some of the expansion. For example, the beginnings of both editions have the same thirteen paragraphs (after the dedication); in the 1800 edition, these thirteen paragraphs run to just under 5.5 pages, while in the 1801 edition, they take up just over 9 pages. The major change is the addition of just over 70 pages of new material that discusses instruments used to convey ideas (the first edition focused primarily on substrates). Once these major changes of words per page and new sections are factored in, we are left with small additions of a line or paragraph here and there. In the case of linen paper, for example, Koops adds a paragraph on page 184, along with a lengthy footnote that ends: "Notwithstanding the flax and hemp plant is dispersed all over Europe, its cultivation is still more proper for the Northern climate" (186). Matthias Koops,

Historical Account of the Substances Which Have Been Used to Describe Events, and to Convey Ideas, from the Earliest Date to the Invention of Paper, 2nd ed (London: Jaques, 1801).

49. Koops, *Substances* (1800), 76.

50. Jacob Christian Schäffer, *Versuche und Muster ohne alle Lumpen oder doch mit einem geringen Zusatze derselben Papier zu machen* (Regensburg, 1765). As the title indicates, Schäffer's paper was strengthened with rag pulp. Many of the paper samples included in the volume I inspected at the Morgan Library (PML 129707) are soft and very fragile; however, the samples do demonstrate the viability of making paper from an amazing assortment of vegetable fibers. See also Weber, *Language of Paper*, 67–68.

51. Koops, *Substances* (1800), 77.

52. Koops (1800), 76–77. Note that in the 1801 edition, Koops emphasizes substitution by changing "and adopted in the manufacture of Paper" to "and adopted *as a substitute for rags* in the manufacture of Paper" (emphasis mine). See Koops, *Substances* (1801), 234.

53. Koops, *Substances* (1800), 15.

54. Koops (1800), 15.

55. Koops (1800), 15–16.

56. Koops (1800), 26, 28. Koops use of "superseded" to narrate book history calls to mind Jonathan Gil Harris's work on untimely matter. We might say, regarding *Substances*, that the straw is explosive, the narrative Koops tells is supersessionary, and the volume on the shelf, beside paper made from rags and paper made from tree pulp, is conjunctive. See Jonathan Gil Harris, *Untimely Matter in the Time of Shakespeare* (Philadelphia: University of Pennsylvania Press, 2009), particularly the introduction, "Palimpsested Time: Toward a Theory of Untimely Matter," 1–25. For a more extended engagement with Harris's ideas about untimeliness, see Chapter 2.

57. Koops, *Substances* (1800), 28–29.

58. Koops (1800), 44.

59. Koops (1800), 44. Koops had already discussed the limited geographical habitat of papyrus, and had pointed out that some believed the plant had gone extinct (he disagreed). See Koops, 34.

60. Koops (1800), 44.

61. Koops (1800), 45.

62. Koops (1800), 49. The designation "Greek parchment" also suggests that the cotton paper was heavily sized with gelatin. In Chapter 4 I discuss animal sizing at length. Although sizing is tangential to the argument of this chapter, it is worth noting that Koops devotes serious attention to sizing practices in *Substances*.

63. As Harris writes, "Supersession appears to guarantee a punctual progression. . . . 'That was then, this is now': the phrase, a neat summation of supersessionary time, presumes an absolute temporal rupture between two self-identical monads, one canceled, the other current" (*Untimely Matter*, 29).

64. Koops, *Substances* (1800), 50.

65. Koops (1800), 51.

66. Koops (1800), 53.

67. Koops (1800), 70. In the second edition, Koops adds the claim that "it is therefore more natural that this invention is to be ascribed, to a country, which was more familiar with linen, and its agriculture, than with the application of cotton" (184–85).

68. Koops, *Substances* (1800), 60.

69. Koops states that one of his primary goals is "that of manufacturing Paper from vegetables, for the purpose of making bank-notes." According to Koops, "because the mixture of vegetables from which the Paper would be made might remain a secret, *if the necessary measures for that purpose are adopted*: consequently no forgery could henceforth be committed on the Bank" (emphasis mine). Koops, *Substances* (1800), 81–82.

70. Koops (1800), 10.

71. Koops (1800), 10.

72. Koops (1800), 76.

73. It is impossible to miss—or leave unacknowledged—the similarities between, on the one hand, eighteenth- and nineteenth-century conversations about rag dependency and alternative paper fibers, and, on the other hand, twentieth- and twenty-first-century conversations about oil dependency and alternative energy sources.

74. Quoted in Hunter, *Papermaking*, 333n. See also Richard Hills, "Some Notes on the Matthias Koops Papers," *The Quarterly: The Review of British Association of Paper Historians*, no. 2 (August 1990): 3–5. Hills also discusses Koops in *Papermaking in Britain*, 133–35.

75. Koops, *Substances* (1800), 78. A footnote keyed to this sentence in Koops reads: "Under the inspection of the truly ingenious Mr. Carpenter." It is the only mention in the volume of Elias Carpenter, Koops's partner, who was possibly as much an innovator and co-inventor, not just an inspector. In the 1801 edition, the phrasing is changed slightly to update the claim, and the reference to Carpenter is dropped; see Koops, *Substances* (1801), 251.

76. Koops, *Substances* (1800), 79.

77. Koops (1800) 80.

78. Koops (1800), 85.

79. That is, in all editions where the appendix is present, it is in fair to good condition. To date, I have examined nine copies of the 1800 edition of Koops at the British Library, the Folger Shakespeare Library, the Morgan Library, the Newberry Library, the New York Public Library, the University of McGill Special Collections, the University of Delaware Special Collections, and the University of Wisconsin Special Collections. All copies I have seen appear to be in good condition. The key features that stand out have to do with chain lines and with paper thickness. In the Newberry copy, and only in that copy out of the nine I have examined, the chain lines in quire X are horizontal (perpendicular to the spine), unlike those in the rest of the book.

80. Koops, *Substances* (1800), 91.

81. R. J. Goulden, "Koops, Matthias (*fl.* 1789–1805)," *Oxford Dictionary of National Biography*, rev. ed., accessed August 3, 2015, http://www.oxforddnb.com/view/article/37644.

82. Keri Davies notes that Koops's paper mill was intended to be "one of the largest industrial enterprises in London," and further suggests it may have inspired William Blake's reference to "Satanic Mills" in his Preface to Milton. Keri Davies, "William Blake and the Straw Paper Manufactory at Millbank," in *Blake in Our Time: Essays in Honour of G. E. Bentley Jr.*, ed. Karen Mulhallen (Toronto: University of Toronto Press, 2010), 235, 233–34.

83. Calling attention to the role of wood pulp in English printing (and, in particular, the North American pine tree), Peter Stoicheff observes that "prior to 1750, approximately one hundred new titles were published annually in England; by 1825 approximately 600 annually; by the end of the nineteenth century approximately 6,000." Peter Stoicheff, "Materials and Meanings," in *The Cambridge Companion to the History of the Book*, ed. Leslie Howsam (Cambridge: Cambridge University Press, 2014), 84.

84. Joel Munsell, *Chronology of the Origin and Progress of Paper and Paper-Making*, 5th ed., with additions (Albany: J. Munsell, 1876), 4. The 1876 edition is the first to use "progress" in the title. At the end of his highly influential and often-cited *Papermaking: The History and Technique of an Ancient Craft* (1947), Hunter acknowledges a debt the work owes to Munsell's *Chronology* (463–64). Munsell, who was both author and printer, published the first edition in 1856, within two years of the first printing of *Walden.* Incidentally, according to Dard Hunter, in 1856, "consumption of paper in the United States had reached a point where it equalled that of England and France combined" (*Papermaking*, 558).

85. On cheapness and natural resources, Wendell Berry argues that "our present willingness to destroy our country is without limit. . . . we have been living under a policy of cheap food—cheap at any cost—founded foursquare upon a policy of cheap energy—cheap at any cost." Wendell Berry, "'Leadership from the Bottom': Wendell Berry on Rural Revival," *New Atlantis*, no. 21 (2008): 128.

86. Taylor, *Hemp-seed*, 24–25; E2v-E3r.

87. John Hales, *A Compendious or Briefe Examination of Certayne Ordinary Complaints* [. . .], STC 23133 (Imprinted at London: In Fleetstreate, neere vnto Saincte Dunstones Church, by Thomas Marshe, 1581), fol. 26a; G2r.

88. Taylor, *Hemp-seed*, 19–20; D4r-v. Taylor asserts that flax and hemp "are male and female, both one, and the same" (unnumbered; B2r).

89. Taylor, *Hemp-seed*, 20; D4v.

90. Taylor, 26–27; E3v-E4r.

91. Adam G. Hooks, *Selling Shakespeare: Biography, Bibliography, and the Book Trade* (Cambridge: Cambridge University Press, 2016), 18. See also Folger Shakespeare Library, "The Praise of Hemp-seed: John Taylor references Shakespeare by name," Shakespeare Documented, May 25, 2017, http://www

.shakespearedocumented.org/exhibition/document/praise-hemp-seed-john-taylor-references-shakespeare-name.

92. Hooks, *Selling Shakespeare*, 18.

93. For a close reading of *Sonnet* 15 and of biblical/botanical links to Shakespeare, see Joshua Calhoun, "Ecosystemic Shakespeare: Vegetable Memorabilia in the Sonnets," *Shakespeare Studies* 39 (2011): 64–73.

94. I reference MA 1302.8, one of the journals kept by Thoreau at Walden Pond and held at the Morgan Library. See also "Journal?: Autograph Manuscript Signed?: Walden [Concord], 1845 July 5–1846 Mar. 27.," The Morgan Library & Museum, accessed July 25, 2017, https://www.themorgan.org/literary-historical/288111.

95. Bland, *Guide*, 25.

96. See Elise Lemire, *Black Walden: Slavery and Its Aftermath in Concord, Massachusetts* (Philadelphia: University of Pennsylvania Press, 2009), 103. White was not her given name, but a name she took later in life when required by the census (136).

97. Lemire, *Black Walden*, 131.

98. Thoreau, *Walden*, 279–80.

99. Lemire, *Black Walden*, 131.

100. For an intriguing take on the effects of the cotton gin both on the liberation of American slaves and on the failure of Matthias Koops's paper mill, see Verner W. Clapp, "The Story of Permanent/Durable Book Paper, 1115–1970," *Restaurator* 1, no. 3 (1972): 1–51.

101. Bland, *Guide*, 29. Heather Wolfe notes the circuitous path of paper imports: "Almost all of England's paper came from France during the Elizabethan and Jacobean periods, some of it directly from Normandy (Rouen), and much of it via Dutch paper merchants in Amsterdam." Heather Wolfe, "How Expensive Was Writing Paper in Early Modern England?" (presentation, Renaissance Society of America Conference, Washington, DC, March 22–24, 2012), 3. See also Heather Wolfe, "Was Early Modern Writing Paper Expensive?," The Collation: Research and Exploration at the Folger, February 13, 2018, https://collation.folger.edu/2018/02/writing-paper-expensive.

102. John Bidwell, "French Paper in English Books," in *The Cambridge History of the Book in Britain*, vol. 4, *1557–1695*, ed. John Barnard and D. F. McKenzie (Cambridge: Cambridge University Press, 2002), 585.

103. D. C. Coleman, *The British Paper Industry, 1495–1860: A Study in Industrial Growth* (Oxford: Clarendon Press, 1958), 53.

CHAPTER 2

Note to epigraph: Timothy Barrett, "Aesthetics and the Future of the Craft," *Hand Papermaking* 11, no. 2 (Winter 1996), 17.

1. Sean Shesgreen, "Canonizing the Canonizer: A Short History of *The Norton Anthology of English Literature*," *Critical Inquiry* 35, no. 2 (2009): 296–97. For a brief review of the "great commotion" that followed the publication of Shesgreen's article, see Tracy J. Prince, *Culture Wars in British Literature: Multiculturalism and National Identity* (Jefferson, NC: McFarland, 2012), 7.

2. Shesgreen, "Canonizing the Canonizer," 297.

3. Shesgreen, 297.

4. Pamela Robinson, "The Format of Books: Books, Booklets, and Rolls," in *The Cambridge History of the Book in Britain*, vol. 2, *1100–1400*, ed. Nigel J. Morgan and Rodney M. Thomson (Cambridge: Cambridge University Press, 2008), 49.

5. Peter S. Hawkins, *Dante's Testaments: Essays in Scriptural Imagination* (Stanford: Stanford University Press 1999), 39.

6. Lucien Febvre and Henri-Jean Martin, *The Coming of the Book: The Impact of Printing, 1450–1800*, ed. Geoffrey Nowell-Smith and David Wooton, trans. David Gerard (London: Verso, 1976), 29–30. According to Febvre and Martin, "there was nothing to prevent" the development of printing in the fourteenth century except "one primary material without which it would have been impossible, namely paper" (29–30). On the relative cheapness of paper to parchment, see R. J. Lyall, "Materials: The Paper Revolution," in *Book Production and Publishing in Britain, 1375–1475*, ed. Jeremy Griffiths and Derek

Pearsall (Cambridge: Cambridge University Press, 1989), 11; Marcel Thomas, "Manuscripts," in Febvre and Martin, *The Coming of the Book*, 16–18; and Allan Stevenson, *The Problem of the Missale Speciale* (London: The Bibliographical Society, 1967), 49–50.

7. Ian Green, *Print and Protestantism in Early Modern England* (Oxford: Oxford University Press, 2000), 553.

8. D. F. McKenzie, *Bibliography and the Sociology of Texts* (1986; repr., Cambridge: Cambridge University Press, 1999), 13.

9. Lori Anne Ferrell similarly claims that the textbook-like, "portable, material, ubiquitous" Paris Bibles were "symbols of a changing, increasingly intimate relationship between the Bible and ordinary people." Lori Anne Ferrell, *The Bible and the People* (New Haven: Yale University Press, 2008), 38.

10. John Dryden, *Religio Laici; or, A Layman's Faith*, Wing D2342 (London: Printed for Jacob Tonson, 1682), 25; E1r.

11. Margreta de Grazia, "Words as Things," *Shakespeare Studies* 28 (2000): 231.

12. De Grazia, 233.

13. Henry Vaughan, "To the Holy Bible," in *Silex Scintillans: Sacred Poems and Private Ejaculations*, 2nd ed., Wing V126 (London: Printed for Henry Crips and Lodowick Lloyd, 1655), 81; H1r, line 16.

14. Defining polychronic (and distinguishing it from multitemporal), Harris writes: "Serres's notion of the polychronic draws on the first, chronological meaning of time in asserting that objects collate many different moments, as suggested by Latour's polytemporal toolbox and genes. By contrast, Serres's notion of the multitemporal evokes the second meaning of time. In its polychronicity, an object can prompt many different understandings and experiences of temporality—that is, of the relations between now and then, old and new, before and after." Jonathan Gil Harris, *Untimely Matter in the Time of Shakespeare* (Philadelphia: University of Pennsylvania Press, 2009), 4.

15. Henry Vaughan, "The Book," in *Silex Scintillans*, 90–81 [*sic*]; G8v-H1r.

16. A[lan] W. Rudrum, "Henry Vaughan's 'The Book'; A Hermetic Poem," *Journal of the Australasian Universities Language and Literature Association* 16 (1961): 162; see also James D. Simmonds, "Vaughan's 'The Book': Hermetic or Meditative?," *Neophilologus* 47 (1963): 320–28.

17. William H. Sherman, *Used Books: Marking Readers in Renaissance England* (Philadelphia: University of Pennsylvania Press, 2008), 71.

18. Sherman, *Used Books*, 72.

19. William Prynne, *Histrio-mastix. The players scourge, or, actores tragœdie*, STC 20464a (London: Printed by E[dward] A[llde, Augustine Mathewes, Thomas Cotes] and W[illiam] I[ones] for Michael Sparke, 1633), **6v.

20. John Vicars, *Unholsome henbane between two fragrant roses*, Wing V332 (London: Printed for John Rothwell, 1645), 7–8; A4r-v.

21. Vicars, *Unholsome henbane*, 8; A4v.

22. Vicars, 8; A4v.

23. On post-Reformation Bible-marking practices, see Sherman, *Used Books*, 71–86. On sinking paper, see Chapter 4, where I discuss animal sizing and sinking paper at length. Similarly, *The Norton Anthology*'s paper "bleeds profusely" when annotated with non-ballpoint pens (Shesgreen, "Canonizing the Canonizer," 297).

24. Edward Leigh, *A treatise of divinity consisting of three books*, Wing L1011 (London: Printed by E. Griffin for William Lee, 1646), 38–41; G3v-H1r.

25. Leigh, *A treatise of divinity*, 41; H1r.

26. Green, *Print and Protestantism*, 98.

27. Green, 100.

28. Green, 56–62.

29. Michael Sparke, *Scintilla, or, a light broken into darke Warehouses*, Wing S4818B (London: Printed, not for profit, but for the common weles good: and no where to be sold, but some where to be given, 1641), A1r.

30. Sparke, *Scintilla*, 2–3; A2v-A3r.

31. Sparke, 1; A2r.

32. Green, *Print and Protestantism*, 39.

33. David McKitterick, *A History of Cambridge University Press*, vol. 1, *Printing and the Book Trade in Cambridge, 1534–1698* (Cambridge: Cambridge University Press, 1992), 286.

34. Jeffrey Todd Knight, "Making Shakespeare's Books: Assembly and Intertextuality in the Archives," *Shakespeare Quarterly* 60, no. 3 (Fall 2009): 309. See also Knight, *Bound to Read: Compilations, Collections, and the Making of Renaissance Literature* (Philadelphia: University of Pennsylvania Press, 2013), where he emphasizes the " 'proximate' or 'material intertextuality' " of these composite volumes (82).

35. Leigh, *A treatise of divinity*, 40; G4v.

36. Colin H. Bloy, *A History of Printing Ink, Balls, and Rollers, 1440–1850* (1967; repr., London: The Wynkyn de Worde Society, 1972), 14.

37. *The Holy Bible Containing the Old Testament and the Neuu / Newly Translated out of the Originall Tongues* [. . .], STC 2296 (Printed at London: By Robert Barker [. . .], and by the assignes of Iohn Bill, 1631). The offending passage appears at Exodus 20:14, D3v.

38. "Whyther the apostles left aught vnwriten, that is of necessyte to be byleued" was, according to More, "that mater, whyche is vndoutely one of the moste specyall poyntes that are in debate bytwene these heretyques and vs." Thomas More, *The Confutation of Tyndale's Answer*, ed. Louis A. Schuster, Richard C. Marius, James P. Lusardi, and Richard J. Schoeck, *The Complete Works of St. Thomas More*, vol. 8, pt. 1 (New Haven: Yale University Press, 1973), 254.

39. Germain P. Marc'hadour and Thomas M. C. Lawler, "Scripture in the *Dialogue*," in *A Dialogue Concerning Heresies*, ed. Thomas M. C. Lawler, Germain P. Marc'hadour, and Richard C. Marius, in *The Complete Works of St. Thomas More*, vol. 6, pt. 2 (New Haven: Yale University Press, 1981), 508.

40. David Daniell claims that More's criticism of Tyndale is fundamentally an objection to his translation of four Greek words (*presbuteros*, *ekklesia*, *agape*, *metanoeo*). David Daniell, *William Tyndale: A Biography* (1994; repr., New Haven: Yale University Press, 2001), 269.

41. More, *A Dialogue Concerning Heresies*, in *The Complete Works of St. Thomas More*, vol. 6, pt. 1, 115.

42. See John Krill, *English Artists' Paper: Renaissance to Regency*, 2nd ed. (New Castle, DE: Oak Knoll P and Winterthur Museum, 2002), 9.

43. George Gascoigne, *A Hundreth sundrie Flowres bounde vp in one small Poesie*, STC 11635 (London: Imprinted [by Henrie Bynneman [and Henry Middleton]] for Richarde Smith, 1573), 426; Gg1v, British Library, C.13.a.16.(1).

44. Krill identifies a shive by its appearance as a "large grid-like 'fiber.' " Krill, *English Artists' Paper*, 7.

45. Baldassare Castiglione, *The Covrtyer of Covnt Baldessar Castilio, Devided into foure Bookes*, STC 4779 (London: By Henry Denham, 1577), C5v-C6r, British Library, 8404.ccc.25.

46. Samuel Rutherford, *Lex, Rex: The Law and the Prince*, Wing R2386 (London: Printed for John Field, 1644) C2r, Fisher Library, University of Toronto, Forbes 00804. I am grateful to Randall McLeod for photographing and sharing this example.

47. In a previously published version of parts of this chapter, I suggested that flax needed to be inhabited and worn out in order to be used in papermaking. Such a claim accurately describes historical papermaking practices but misrepresents possibilities; as I show in the previous chapter, paper might be made directly from flax plants instead of linen rags. It is worth noting, however, that paper made from worn-out linen is different and better than paper made directly from flax (an insight I owe to the expertise and experience of Robert Possehl).

48. Robert Hooke, *Micrographia, or, Some Physiological Descriptions of Minute Bodies Made by Magnifying Glasses with Observations and Inquiries thereupon*, Wing H2620 (London: Printed by Jo. Martyn and Ja. Allestry, 1665), 209–10; Ff1r-v.

49. John Carter and Nicolas Barker, *ABC for Book Collectors*, 8th ed. (Newcastle, DE: Oak Knoll Press, 2004), 232.

50. S. Blair Hedges, "Wormholes Record Species History in Space and Time," *Biology Letters* 9 (2012): 1. The article cited focuses primarily on wormholes in wood blocks.

51. Peter Stallybrass, "Introduction to 360 Analysis" (presentation, Book 360 Symposium, Weston Library, Bodleian, Oxford, May 11–12, 2017); see https://sites.google.com/palaeome.org/lvcas-360/home.

52. Hedges, "Wormholes," 1.

53. "Crane.com," Crane and Co., accessed August 15, 2017, http://www.crane.com/about-us/crane-currency.

54. Febvre and Martin, *The Coming of the Book*, 37.

55. Febvre and Martin, 31. On England's struggle to establish domestic paper production, see John Bidwell, "French Paper in English Books," in *The Cambridge History of the Book in Britain*, vol. 2, *1557–1695*, ed. John Barnard and D. F. McKenzie (Cambridge: Cambridge University Press, 2002), 584–87.

56. On human waste as fertilizer for cultivated plants, see Emily Cockayne, *Hubbub: Filth, Noise, and Stench in England, 1600–1770* (New Haven: Yale University Press, 2007), 184.

57. Ann Rosalind Jones and Peter Stallybrass, *Renaissance Clothing and the Materials of Memory* (Cambridge: Cambridge University Press, 2000).

58. John Donne, "A Fvneral Elegie," in *An Anatomy of the Vvorld*, STC 7022 (London: Printed [by William Stansby] for Samuel Macham, 1611), B7r, lines 11–12.

59. John Donne, *Letters to Severall Persons of Honour: Written By John Donne*, Wing D1864 (London: Printed by J. Flesher for Richard Marriot, 1651), 88; M4v.

60. Donne, *Letters*, 224; Ff4v; and 136; S4v.

61. Abraham Cowley, *The Guardian, A Comedie Acted before Prince Charls, His Highness at Trinity-Colledg in Cambridge, upon the twelfth of March, 1641*, Wing C6673 (London: Printed for John Holden, 1650), B1v.

62. Thomas Dekker and George Wilkins, *Iests to make you Merie: With The Coniuring vp of Cock VVatt, (the walking Spirit of Newgate) To tell Tales*, STC 6541 (London: By N[icholas] O[kes] for Nathaniell Butter, 1607), 61; H3r.

63. Jean Baptiste van Helmont, *Van Helmont's Works: Containing his most Excellent Philosophy, Physick, Chirurgery, Anatomy*, trans. J. C., Wing H1397 (London: Printed by Lodowick Lloyd, 1664), 1154; Ggggggg3v.

64. John Tatham, *The Scots Figgaries: or, A Knot of Knaves*, Wing T235 (London: Printed by W. H. for Iohn Tey, 1652), 31; E4r.

65. Juan Luis Vives, *The Instruction of a Christen Woman*, ed. Virginia Walcott Beauchamp, Elizabeth H. Hageman, and Margaret Mikesell (Urbana: University of Illinois Press, 2002), 16–17.

66. Margaret Cavendish, "Letter CL," in *CCXI. Sociable Letters*, Wing N872 (London: Printed by William Wilson, 1664), 314–15; Rr1v-Rr2r.

67. Fiona McNeill, *Poor Women in Shakespeare* (Cambridge: Cambridge University Press, 2007), 33.

68. McNeill, *Poor Women in Shakespeare*, 33–34.

69. Michael Maier, *Lusus Serius: or, Serious Passe-time. A Philosophicall Discourse concerning the Superiority of Creatures under Man*, trans. J. de la Salle [John Hall], Wing M286 (London: Printed for Humphrey Moseley and Tho. Heath, 1654), 138–39; G9v-G10r, Folger Shakespeare Library, 156–125q.

70. Maier, *Lusus Serius*, 81; D12v.

71. Maier, 81–73 [*sic*]; D12v-E1v.

72. Maier, 73–74; E1r-v.

73. See Carole Shammas, "How Self-Sufficient Was Early America?," *Journal of Interdisciplinary History* 13, no. 2 (1982): 254.

74. Maier, *Lusus Serius*, 73; E1r.

75. Maier, 82; E5v; and 82–83; E5v-E6r.

76. Maier, 83; E6r.

77. Maier, 84; E6v.

78. Maier, 84; E6v.

79. Maier, 85; E7r.

80. Maier, 88; E8v.

81. Maier, 129; G5r.

82. Francis Ernest Hutchinson, *Henry Vaughan: A Life and Interpretation* (Oxford: Clarendon Press, 1947), 17–19.

83. Maier, *Lusus Serius*, numbers 102 and 169, unnumbered pages; 2A5r, 2A8r, Folger Shakespeare Library, 156–125q.

84. Harris, *Untimely Matter*, 13.

85. Vaughan, "To the Holy Bible," in *Silex Scintillans*, 81; H1r, lines 5–8 and 16.

86. See Georgia B. Christopher, "In Arcadia, Calvin . . . : A Study of Nature in Henry Vaughan," in *Essential Articles for the Study of Henry Vaughan*, ed. Alan Rudrum (Hamden, CT: Archon Books, 1987), 181; Barbara Kiefer Lewalski, *Protestant Poetics and the Seventeenth-Century Religious Lyric* (Princeton: Princeton University Press, 1979), 346–47; Matthew Prineas, "The Dream of the Book and the Poetry of Failure in Henry Vaughan's *Silex Scintillans*," *English Literary Renaissance* 26, no. 2 (1996): 344, 355; James D. Simmonds, *Masques of God: Form and Theme in the Poetry of Henry Vaughan* (Pittsburgh: University of Pittsburgh Press, 1972), 161–62. See also Diane Kelsey McColley, *Poetry and Ecology in the Age of Milton and Marvell* (Burlington, VT: Ashgate, 2007), 114–15; and Alan Rudrum, "Review: [untitled]," *Review of English Studies* 43, no. 171 (1992): 426 (Rudrum disagrees, but does not offer a counter-argument). Bolstering arguments in support of such a reading, the physical relationship of the two poems creates an interface between the "aged book" described beneath the header "The Book" and the old, used Bible addressed as "O Book!" in the facing, opening lines of "To the Holy Bible."

87. Alan Rudrum, ed., *Henry Vaughan: The Complete Poems* (1976; repr., New Haven: Yale University Press, 1981), 641.

88. McColley, *Poetry and Ecology*, 113.

89. Harris, *Untimely Matter*, 19–20.

90. McKenzie, *Sociology of Texts*, 13.

91. Bruce Holsinger draws attention to what might be a source—or at least an analogue—for Vaughan's skin-covered book boards: a Bible riddle from the tenth-century Exeter Book in which the Bible describes its own creation. Holsinger notes that "once written, the book is bound, in a fascinating description that reverses the speaker's identity: 'shieldboards clothed me and stretched hide, a skin laced with gold.' The book is now speaking as the 'clothed' subject, covered with 'stretched hide' and 'skin': the animal parts needed to enclose the writing." Bruce Holsinger, "Of Pigs and Parchment: Medieval Studies and the Coming of the Animal," *PMLA* 124, no. 2 (March 2009): 622.

92. M. M. Mahood, "Vaughan: The Symphony of Nature," in Rudrum, *Essential Articles for the Study of Henry Vaughan*, 39–40.

93. See also this book's Introduction for further discussion of the ecology of texts as an extension of McKenzie's sociology of texts and, more broadly, of the sociological turn in book history scholarship.

94. McKenzie, *Sociology of Texts*, 12, 29.

95. The word *papyr*, emphasized in italics, recalls the material history of the Bible; for a brief, visual introduction, see the online exhibit *From Papyri to King James: The Transmission of the English Bible*, University of Michigan Library, 2004, https://www.lib.umich.edu/files/collections/papyrus/exhibits/papyri_james/review.html. Here Vaughan seems to be merging "corn of wheat" from John 12:24 with the parable of the wheat and the tares from Matthew 13:24–30.

96. Jones and Stallybrass, *Renaissance Clothing*, 11.

97. Jones and Stallybrass, 14.

98. Simmonds, *Masques of God*, 162–63.

99. Simmonds, "Book," 323.

100. Simmonds, *Masques*, 161.

101. E. C. Pettet, *Of Paradise and Light: A Study of Vaughan's "Silex Scintillans"* (Cambridge: Cambridge University Press, 1960), 32.

102. Christopher, "In Arcadia, Calvin," 170.

CHAPTER 3

Note to epigraph: Lisa Gitelman, *Always Already New: Media, History, and the Data of Culture* (Cambridge, MA: MIT Press, 2008), 6.

1. The blot is 85 mm high x 100 mm wide at its outer points with the volume open as wide as the binding naturally permits. William Shakespeare, *Mr. VVilliam Shakespeares Comedies, Histories, & Tragedies: Published according to the True Originall Copies*, ed. Iohn Heminge and Henry Condell, STC 22273 (London: Printed by Isaac Iaggard, and Ed. Blount, 1623), 96–97; gg5v-gg6r, Folio no. 47, Folger Shakespeare Library, Washington, DC.

2. In the opening scene of *1 Henry IV*, King Henry IV compares Prince Harry with Harry Hotspur. While Hotspur is "the theme of Honor's tongue," the King "See[s] riot and dishonor stain the brow / Of my young Harry." See 1.1.80, 84–85.

3. Francis Quarles, *Enchyridion: containing institutions divine contemplative practicall, morall ethycall oeconomicall politicall,* STC 20543 (London: Tho. Cotes, 1640), Chap. LVIII, sig. H11r.

4. W. W. Greg, one of the most notable among New Bibliographers, on the important "distinction between the significant, or as I shall call them 'substantive', readings of the text, those namely that affect the author's meaning or the essence of his expression, and others, such in general as spelling, punctuation, word-division, and the like, affecting mainly its formal presentation, which may be regarded as the accidents, or as I shall call them 'accidentals', of the text." W. W. Greg, "The Rationale of Copy-Text," *Studies in Bibliography* 3 (1950): 21.

5. D. F. McKenzie, "Typography and Meaning: The Case of William Congreve," in *Making Meaning: "Printers of the Mind" and Other Essays,* ed. Peter D. McDonald and Michael F. Suarez (Amherst: University of Massachusetts Press, 2002), 202.

6. Rita Felski, *Uses of Literature* (Hoboken, NJ: Wiley-Blackwell, 2008), 4.

7. Felski, *Uses of Literature,* 18–21.

8. Anon., *The Famovs Victories of Henry the fifth: Containing the Honourable Battell of Agin-court: As it was plaide by the Queenes Maiesties Players,* STC 13072 (London: Printed by Thomas Creede, 1598), B3v.

9. I cite these lines directly from Folio 47; in the standard text cited throughout this chapter, the lines appear at 5.2.64–73.

10. Of this scene in *Henry V,* Hannibal Hamlin writes, "Henry's strategy in trapping the traitors is exactly that of the prophet Nathan in trapping King David into an admission of his sin." Hannibal Hamlin, *The Bible in Shakespeare* (Oxford: Oxford University Press, 2013), 256n60. In 2 Samuel 12, Nathan famously asks David's judgment on another's offense. When David recommends death for the guilty party, Nathan turns the judgment back on him. Similarly, King Henry solicits the opinion of the three traitors in regard to the offense of one who, supposedly, got drunk and railed against the king. The traitors recommend death for the man. In this instance though, instead of proclaiming the sentence back, Henry gives each man a paper that condemns them. As they read, he remarks: "What see you in those papers, that you lose / So much complexion?—Look you, how they change. / Their cheeks are paper" (2.2.77–79).

11. In *Richard III,* the title character equates "blots and stains" and suggests they are marks that can be acquitted (literally, paid for as if they are debts):

> But if black scandal or foul-faced reproach
> Attend the sequel of your imposition,
> Your mere enforcement shall acquittance me
> From all the impure blots and stains thereof
> (3.7.233–36)

12. Barbara Hodgdon, "Historiography and the Uses of History," in *The First Part of King Henry the Fourth: Texts and Contexts,* ed. Barbara Hodgdon (Boston: Bedford Books, 1997), 121.

13. H. Austin Whitver, "Materiality of Memory in Shakespeare's Second Tetralogy," *Studies in English Literature* 56, no. 2 (Spring 2016): 296. See also Paola Pugliatti, who claims that in Shakespeare's history plays we find a "more mature form of historiography, one conscious that the past, apparently crystallized in unchanging documents and monuments, may prove unstable and multiform." Paolka Pugliatti, "'More than history can pattern': The Jack Cade Rebellion in Shakespeare's *Henry VI,*" *Journal of Medieval and Renaissance Studies* 22, no. 3 (Fall 1992): 454.

14. A more diplomatic transcription of these words as they appear on the page in question would be "wash'd," "forgotten," "Remission," "conuert," "vnstained," and "race."

15. Randall McLeod, "Obliterature: Reading a Censored Text of Donne's 'To his mistress going to bed,'" *English Manuscript Studies* 12 (2005): 83. McLeod is quoting Gary A. Stringer, ed., *The Variorum Edition of the Poetry of John Donne,* vol. 2, *The Elegies* (Bloomington: Indiana University Press, 2000), 175, 178.

16. McLeod, "Obliterature," 88.

17. McLeod, 122.

18. McLeod is also notably attentive to the relationship between literary form and critical desire in the making and marring of Donne's "To his mistress going to bed." Emotion, he argues, not just textual

science, is at play in both Gardner's edition (1912) and Grierson's edition (1965) of the poem: "It seems to have made a difference to both of them," McLeod suggests, "whether the sex in the offing is a nuptial celebration or just raw fucking." McLeod, "Obliterature," 102.

19. Mark Bland, *A Guide to Early Printed Books and Manuscripts* (Hoboken, NJ: Wiley, 2010), 18.

20. Wendy Wall, "Reading for the Blot: Textual Desire in Early Modern England," in *Reading and Writing in Shakespeare*, ed. David M. Bergeron (Newark: University of Delaware Press, 1996), 132.

21. Jonathan Walker, "Reading Materiality: The Literary Critical Treatment of Physical Texts," *Renaissance Drama* 41, nos. 1–2 (Fall 2013): 199.

22. Walker, "Reading Materiality," 232, 201.

23. See *Macbeth*, 5.5.24 ("recorded time") and *2 Henry VI*, 4.7.13 ("records of the realm").

24. That is, *Richard II*, *1* and *2 Henry IV*, and *Henry V*, also referred to as the Second Tetralogy. The First Tetralogy, so named because Shakespeare wrote these plays first, and despite the fact that it dramatizes the reigns of later kings, includes *1*, *2*, and *3 Henry VI* and *Richard III*.

25. For the purposes of my textual frequency analysis, I utilized the Folger Digital Texts edition of Shakespeare's collected works. Unlike other digital Shakespeare collections, this online resource provides a modern and scholarly edition that can be downloaded in multiple formats for various sorts of textual analysis or manipulation. For this project, I consulted the text-only files rather than the application programming interface (API) so I could understand the context of the words in question and more easily search synonyms or words that share the same root (memory, memorandum, remember). Other online Shakespeare collections utilize outdated editions, like Open Source Shakespeare, which uses the 1864 Globe Edition (https://www.opensourceshakespeare.org/). Some readily available online editions provide full, but unreliable Shakespeare texts (Project Gutenberg) or use interfaces that prevent advanced textual analysis (MIT, *The Complete Works of Shakespeare*, http://shakespeare.mit.edu/).

26. Falstaff's "stained with travel and sweating with desire to see him" (5.5.24–25) is the closest thing in the play to "blot." There, too, the bad (stains) are meant to record the good: Falstaff's willingness "to ride day and night" (5.5.20), without concern for himself, so that he can be at the new king's procession.

27. Robert N. Watson, *Back to Nature: The Green and the Real in the Late Renaissance* (Philadelphia: University of Pennsylvania Press, 2005), 79.

28. The Folio remedies the problem, expurgating profanity from the text in accordance with the Act of 1606, which restricted profanity on stage (but not necessarily on the page). See Grace Ioppolo, *Revising Shakespeare* (Cambridge, MA: Harvard University Press, 1991), 78–81.

29. Catalogue entry for Folger shelf mark STC 22288, *Hamnet: Folger Library Catalogue*, accessed December 6, 2018, hamnet.folger.edu (emphasis mine); William Shakespeare et al., *The Second Part of Henrie the Fourth*, STC 22288 (London: Printed by V[alentine]. S[immes] for Andrew Wise, and William Aspley, 1600).

30. There is another Folger Folio that is well known for its many blots: the St. Alban's Second Folio (STC 22274, copy 7), a censored copy that, according to the catalogue, "bears (on title page) the certificate of a censor for the Inquisition." In that volume, "Masse" in "By the Masse" is censored from *1 Henry IV* (58; e4v), but "Iesu" and "Iesus" in *2 Henry IV* had already been censored—we might say obliterated—prior to printing. Catalogue entry for Folger shelf mark STC 22274 Fo.2 no.07, *Hamnet: Folger Library Catalogue*, accessed December 6, 2018, hamnet.folger.edu.

31. On the chronology of the play texts, actual and imagined, see Adam G. Hooks, "Making Histories; or, Shakespeare's *Ring*," in *The Book in History, the Book as History: New Intersections of the Material Text*, ed. Heidi Brayman, Jesse M. Lander, and Zachary Lesser (New Haven: Beinecke Rare Book & Manuscript Library, Yale University, 2016), 341–73. The opening Chorus does not appear in any of the quarto editions of the play.

32. On use of the feminine pronoun for Rumor, see Meredith Evans, "Rumor, the Breath of Kings, and the Body of Law in *2 Henry IV*," *Shakespeare Quarterly* 60, no. 1 (Spring 2009): 1–2.

33. Northumberland had claimed sickness for his absence at Shrewsbury, leaving his son grossly outnumbered. Here Rumor pulls the audience close by intimating to us that Northumberland is "crafty-sick"; we suspect that Northumberland's sickness in *1 Henry IV* was a convenient excuse, but we have no way of knowing for certain.

34. Loren M. Blinde, "Rumored History in Shakespeare's *2 Henry IV*," *English Literary Renaissance* 38, no. 1 (Winter 2008): 44. See also Blinde's suggestion that "by presenting Rumor as our historian,

Shakespeare encourages the audience members to include a fundamental sense of unreliability in their thinking about history" (35).

35. On erasable tables, see Peter Stallybrass, Roger Chartier, John Franklin Mowery, and Heather Wolfe, "Hamlet's Tables and the Technologies of Writing in Renaissance England," *Shakespeare Quarterly* 55, no. 4 (Winter 2004): 379–419.

36. Even Bardolph's name gestures toward confusion and error. Whereas Lord Bardolph is historically based on Northumberland's ally Thomas, 5th Baron of Bardolf, Lord of Wormegay, Falstaff's manservant Bardolph seems to be a composite character. Before the first quarto of *1 Henry IV* was printed, he was Russell (to Falstaff's Oldcastle). Even in the first quarto of *1 Henry IV* he appears as "Bardol" or "Bardoll"; Quarto A of *2 Henry IV* conflates his name with Lord Bardolph, and thereafter his name was printed as Bardolph in *Henry V*, *Merry Wives of Windsor*, and then the Folio version of *1 Henry IV*; see James C. Bulman, ed., *King Henry IV Part 2*, by William Shakespeare (London: Bloomsbury Arden Shakespeare, 2016), 155n12 and 157n26.

37. Elsewhere in the play, Henry IV will send his Page to deliver not an oral message, but pages containing a written message (3.1.1ff.).

38. Lois Potter, "The Second Tetralogy: Performance as Interpretation," in *A Companion to Shakespeare's Works*, ed. Richard Dutton and Jean Howard (Malden, MA: Blackwell, 2003), 2:289.

39. Eulogizing her husband, Hotspur, Lady Percy claims she "never shall have length of life enough / To rain upon remembrance with mine eyes / That it may grow and sprout as high as heaven / For recordation to my noble husband" (2.3.61–64).

40. *2 Henry IV*, 3.1, the scene in which King Henry IV quotes lines he should not have known from *Richard II*, does not appear in Quarto A; it appears in Quarto B and is included in the First Folio. Many modern editions render a blended text based on Quarto A but including 3.1 as well as the Folio-only passages. René Weis offers a usefully detailed review of the texts in the Oxford World's Classics edition of the play. See René Weis, "The Texts of *2 Henry IV*," in *The Oxford Shakespeare: Henry IV Part Two*, by William Shakespeare, ed. René Weis (Oxford: Oxford University Press, 1997), 78–100.

41. Notably, neither *Richard II* nor *1 Henry IV* has a character named Warwick (whom King Henry calls "Nevil" in the lines above), so even the king's memory of who was in his service is in question.

42. Bulman, *Henry IV Part Two*, 289n.

43. Peter Beal, *A Dictionary of English Manuscript Terminology, 1450–2000* (Oxford: Oxford University Press, 2008), 169. The actual, and more frequently cited, name, even in the Renaissance, was Galtres Forest or the Forest of Galtres.

44. Stallybrass, Chartier, Mowery, and Wolfe, "Hamlet's Tables," 418, 417.

45. I am grateful to Karen Britland for pointing out how crucial this topic is in Phillip Massinger's *A New Way to Pay Old Debts*, a play in which the erasure of a parchment deed plays a key role in the plot's development and ultimate resolution.

46. Beal, *A Dictionary*, 283.

47. Michael Finlay claims "vellum" should be "reserved for the parchment made from the skin of a calf, which is of high quality." Finlay also notes, however, that the high quality of vellum and the lower quality of sheepskin—especially sheepskin that has been split to create two sheets—eventually caused other kinds of high-quality parchment to be called vellum. Michael Finlay, *Western Writing Implements in the Age of the Quill Pen* (Carlisle, UK: Plains Books, 1990), 29.

48. See, for instance, Ruth A. Johnston, *All Things Medieval: An Encyclopedia of the Medieval World*, vol. 1, A-I (Santa Barbara: Greenwood, 2011), 74.

49. Finlay, *Western Writing Implements*, 13–14, and see figs. 41–44 on pp. 103–4. See also Joe Nickell, *Pen, Ink, & Evidence: A Study of Writing and Writing Materials for the Penman, Collector, and Document Detective* (New Castle, DE: Oak Knoll Press, 2000), 64, and see fig. 7.14 and caption on p. 65.

50. However, Nicole Flibbert has found recent evidence of knife erasures on heavily sized paper (see Chapter 4 for a full discussion of sizing). Nicole Flibbert, "The 'Art of Scraping': Knife Erasures in Seventeenth-Century English Manuscript Plays," *Journal of the Penn Manuscript Collective* 1, no. 1 (August 1, 2017), https://repository.upenn.edu/manuscript_collective_journal/vol1/iss1/3.

51. In addition to razing and scraping, rubbing is often cited as a means of correction. See, for example, Chaucer's complaint to his scrivener Adam: "So ofte adaye I mot thy werk renewe, / It to correcte and eke to rubbe and scrape." Geoffrey Chaucer, *The Riverside Chaucer*, ed. Larry Dean Benson (New York: Oxford University Press, 2008), 650, lines 5–6.

52. John Milton, *Of Reformation touching church-discipline in England and the causes that hitherto have hindred it*, Wing M2134 ([London]: Printed for Thomas Underhill, 1641), 25; E1r.

53. Michelle Brown and Patricia Lovett, *The Historical Source Book for Scribes* (Toronto: University of Toronto Press, 1999), 14.

54. The speaker in the dialogue further explains that if anyone should "err in writing the roll, either in the name or the amount or in the matter in which the greater force of the writing consists, he shall not presume to erase, but shall cancel by drawing, underneath, a thin line, and shall write directly following what is necessary." See "V. What Is the Office of the President, and of All Those Who Sit There Officially: As to the Scribe of the Treasurer," in *The Dialogue Concerning the Exchequer, circa 1180*, from *The Avalon Project: Documents in Law, History, and Diplomacy*, Yale Law School Lillian Goldman Law Library, accessed December 6, 2018, http://avalon.law.yale.edu/medieval/excheq.asp. *The Avalon Project*'s source is Ernest F. Henderson, *Select Historical Documents of the Middle Ages* (London: George Bell and Sons, 1896). Bioarchaeologist Matthew Collins explains, "Sheepskin parchments will delaminate, due to the greater density of hair follicles, and consequently of more sebaceous glands, which breaks the dense collagen fibrils of the skins" (Matthew Collins and Peter Stallybrass, email correspondence with the author).

55. L. C. Hector, *The Handwriting of English Documents*, 2nd ed. (London: Edward Arnold, 1966), 19–20.

56. Nickell, *Pen, Ink, & Evidence*, 35.

57. John Haines, "Erasures in Thirteenth-Century Music," in *Music and Medieval Manuscripts: Paleography and Performance; Essays Dedicated to Andrew Hughes*, ed. Andrew Hughes, John Dickinson Haines, and Randall Rosenfeld (Burlington, VT: Ashgate, 2004), 63–64. Haines posits that the *Jena* manuscript was created in the 1320s to 1340s.

58. H. Wayne Storey, "Doubting Petrarca's Last Words: Erasure in MS Vaticano Latino 3195," in *Petrarch and the Textual Origins of Interpretation*, ed. Teodolinda Barolini and H. Wayne Storey (Boston: Brill, 2007), 74.

59. Bernard J. Muir, "Editing the Exeter Book: A Progress Report," in *Medieval Texts and Images: Studies of Manuscripts from the Middle Ages*, ed. Margaret M. Manion and Bernard J. Muir (Chur: Harwood Academic, 1991), 150.

60. Muir, "Editing the Exeter Book," 156–57.

61. Sarah Kay, "Original Skin: Flaying, Reading, and Thinking in the Legend of Saint Bartholomew and Other Works," *Journal of Medieval and Early Modern Studies* 36, no. 1 (Winter 2006): 35. See also Bruce Holsinger, "Of Pigs and Parchment: Medieval Studies and the Coming of the Animal," *PMLA* 124, no. 2 (March 2009): 616–23.

62. Haines, "Erasures in Thirteenth-Century Music," 66.

63. D. C. Coleman, *The British Paper Industry, 1495–1860: A Study in Industrial Growth* (Oxford: Clarendon Press, 1958), 10.

64. Hector, *The Handwriting of English Documents*, 17. Papermaking itself was not established in England until the end of the fifteenth century—John Tate's paper mill, which was visited by Henry VII—but this venture ultimately failed. In fact, although the end of the sixteenth century saw a rise in English papermaking, the making of white paper in England did not take a permanent hold until the end of the seventeenth century, after the establishment of the Company of White-Paper-Makers and the "Paper Act" of 1690.

65. M. T. Clanchy, "Parchment and Paper Manuscript Culture, 1100–1500," in *A Companion to the History of the Book*, ed. Jonathan Rose and Simon Eliot (Chichester, West Sussex: Wiley-Blackwell, 2009), 196.

66. Ann Blair, "The Rise of Note-Taking in Early Modern Europe," *Intellectual History Review* 20, no. 3 (2010): 310.

67. Blair, "The Rise of Note-Taking," 309.

68. Coleman, *The British Paper Industry*, 10.

69. Coleman, 10.

70. See Stallybrass, Chartier, Mowery, and Wolfe, "Hamlet's Tables," 409.

71. Alan Stewart, *Shakespeare's Letters* (Oxford: Oxford University Press, 2008), 41.

72. *OED Online*, s.vv. "blot, n.1," "† beblot, v," March 2019, http://www.oed.com. See also *OED Online*, s.v. "blot, v," http://www.oed.com. The *OED*'s earliest date for "blot" as a noun (Chaucer, 1386) is metaphorical or figurative: "A moral stain; a disgrace, fault, blemish."

73. The Geneva Bible used some form of the word just three times in the Old Testament, one time in the New Testament, and zero times in the Apocrypha. In the KJB, some form of "blot" appears eighteen times in the Old Testament, three times in the New Testament, and eight times in the Apocrypha.

74. See Numbers 5:23, Job 31:7, Proverbs 9:7, and 2 Peter 2:13.

75. *The Holy Bible*, STC 2216 (London: By Robert Barker, 1611).

76. James Strong, *The New Strong's Exhaustive Concordance of the Bible* (Cambridge, ON: Thomas Nelson, 1984), 64, no. 4229.

77. For the sake of comparison, the Latin Vulgate has "dele me de libro tuo."

78. The list of alternate words used includes "put out," "destroy," "wipe," "wipe away," and "abolish."

79. The Hebrew word is *çêpher* (transliterated) and most precisely indicates a writing or document. The word appears nearly two hundred times in the Old Testament, and in one instance, Isaiah 34:4, the KJB translates it as "scroll."

80. Genesis 7:4.

81. Hannibal Hamlin, *Psalm Culture and Early Modern English Literature* (Cambridge: Cambridge University Press, 2004), 173–74. Hamlin devotes an entire chapter to Psalm 51.

82. Hamlin, *Psalm Culture*, 174.

83. The verse begins "Miserere mei Deus" and the verb "dele"—root of the English "delete"—is used to indicate that action that Geneva translates as "put awaie." In the Latin Vulgate, which follows the Septuagint rather than the Masoretic numbering of the Psalms, Psalm 51 appears as Psalm 50 (and the verse in question is actually Psalm 50:3 because the headnote is numbered as 1–2).

84. Ian Donaldson, "Jonson, Benjamin [Ben] (1572–1637), poet and playwright," *Oxford Dictionary of National Biography*, accessed July 21, 2017, http://www.oxforddnb.com/view/article/15116.

85. *The Geneva Bible: A Facsimile of the 1560 Edition* (Peabody, MA: Hendrickson, 2007). fol. 245v; Qqiv.

86. In a fascinating article on illustrations of the Penitential Psalms in medieval and Renaissance Books of Hours, Clare L. Costley explains that "Books of Hours, commonly known as primers in England, were produced in vast quantity from the mid-thirteenth century." They were typically illustrated, and the image that appeared most frequently was King David repenting for his sins. "But the images for the Penitential Psalms in Books of Hours created from the 1480s testify to a remarkable shift in focus. . . . they call attention to David's sin (rather than his repentance), portraying the moment when the king first catches sight of Uriah's wife," Costley observes. With the help of twenty images, the article convincingly illustrates what Costley describes as "a shift in attitudes toward penance and sin at the end of the Middle Ages." The chronological overlay of the shift that Costley observes in representations of David with the technological shift from parchment to paper, from erasure to blotting, is particularly intriguing and, I think, worthy of further consideration. See Clare L. Costley, "David, Bathsheba, and the Penitential Psalms," *Renaissance Quarterly* 57, no. 4 (Winter 2004): 1244, 1235.

87. George Gascoigne, *A Hundreth Sundrie Flowres*, ed. G. W. Pigman III (Oxford: Clarendon Press, 2000), 260–61. Pigman claims Gascoigne has a precedent from Ovid for citing David in this context (625n46).

88. Hamlin, *Psalm Culture*, 191.

89. European Commission, "Data Protection in the EU, Article 17: 'Right to be forgotten and to erasure,'" *Proposal for a Regulation of the European Parliament and of the Council on the protection of individuals with regard to the processing of personal data and on the free movement of such data (General Data Protection Regulation)*, accessed June 14, 2018, https://ec.europa.eu/info/law/law-topic/data-protection/data-protection-eu_en; see also Meg Leta Jones, *Ctrl + Z: The Right to Be Forgotten* (New York: NYU Press, 2018), 9–11.

CHAPTER 4

Note to epigraph: Simon Barcham Green, "Sizing at Hayle Mill," *Hand Papermaking* 30, no. 2 (Winter 2015): 11.

1. Jean Imberdis, *Papyrus Sive Ars Conficiendae Papyri (Papyrus, or The Craft of Paper)*, trans. Eric Laughton (Hilversum, Holland: The Paper Publications Society, 1952), 59. Imberdis's 486-line Latin poem was first printed in 1693.

2. Ian Hutchby, "Technologies, Texts, and Affordances," *Sociology* 35, no. 2 (2001): 453. See also Hutchby's incitement to "pay more attention to the material substratum which underpins the very possibility of different courses of action in relation to an artefact; and which frames the practices through which technologies come to be involved in the weave of ordinary conduct" (450).

3. Timothy Barrett, "Sizing Terminology and Practice," *Hand Papermaking* 30, no. 2 (Winter 2015): 3. Barrett credits Ray Janes, his professor from Western Michigan University's School of Paper Science and Engineering, for this definition.

4. Michael Twyman, *The British Library Guide to Printing* (Toronto: University of Toronto Press, 1999), 13.

5. Joe Nickell writes, "To be used for writing . . . paper required a harder surface to minimize its absorbency; otherwise the ink would spread among the fibers." Joe Nickell, *Pen, Ink, & Evidence* (New Castle, DE: Oak Knoll Press, 2000), 88.

6. William H. Sherman, *Used Books: Marking Readers in Renaissance England* (Philadelphia: University of Pennsylvania Press, 2008), 5, 188n8.

7. Sherman, *Used Books*, 5.

8. See Timothy Barrett et al., Paper Through Time: Nondestructive Analysis of 14th- Through 19th-Century Papers, The University of Iowa, last modified July 31, 2018, http://paper.lib.uiowa.edu. As Barrett and his collaborators have demonstrated, gelatin sizing is largely responsible for the exceptional durability and stability of early handmade papers. On the project website, Barrett also offers a particularly extensive description of sizing practices in his article "European Papermaking Techniques, 1300–1800," Paper Through Time.

9. Imberdis, *Papyrus*, 59–61.

10. See descriptions of the angle of contact method and the Cobb test in John Baty, "Sizing Science: Can Physical Science and Engineering Answer Some of Our Lingering Questions?," *Hand Papermaking* 30, no. 2 (Winter 2015): 9–10. See also Timothy Barrett, "Listening to Paper," *Hand Papermaking* 32, no. 2 (Winter 2017): 8.

11. *The Holy Bible [The Geneva Bible]*, STC 2129 (London: By Christopher Barker, printer to the Queenes most excellent Maiestie, 1580), Folger shelf mark STC 2129, "The Newby Bible" (1580). For a fuller description of this Bible as evidence of "discontinuous reading," see Peter Stallybrass, "Books and Scrolls: Navigating the Bible," in *Books and Readers in Early Modern England*, ed. Jennifer Andersen and Elizabeth Sauer (Philadelphia: University of Pennsylvania Press, 2001), 46, 51ff.

12. Katherine Acheson, "The Occupation of the Margins: Writing, Space, and Early Modern Women," in *Early Modern English Marginalia*, ed. Katherine Acheson, Material Readings in Early Modern Culture (London: Routledge, 2018), 80tk.

13. *The Holy Bible [The Geneva Bible]*, Ii3r.

14. See also Alan Stewart, *Shakespeare's Letters* (Oxford: Oxford University Press, 2008), 44. Stewart calls attention both to paper's "irritating ability to absorb ink" and to the language of sinking used by Renaissance writers.

15. John Vicars, *Unholsome henbane between two fragrant roses*, Wing V332 (London: Printed for John Rothwell, 1645), 8; A4v. See also my discussion of this passage in Chapter 2.

16. Nathaniel Shute, *Corona Charitatis, The Crowne of Charitie: A Sermon Preacht in Mercers Chappell, May 10. 1625*, STC 22466 (London: Printed by W. Stansby for Samuel Man, dwelling at the Swanne in Pauls Church-yard, 1626), 10; C1v.

17. D. R., *Naaman the Syrian His Disease and Cvre Discovering Lively to the Reader the Spirituall Leprosie of Sinne and Selfe-Love*, Wing R1799 (London: Printed by Th. Harper for Philip Nevil, 1642), 202; R4v.

18. See Andrew Pettegree, *The Book in the Renaissance* (New Haven: Yale University Press, 2011), xiv; on the "value of diversification," see also 97.

19. William Horman, *Vulgaria Uiri Doctissimi Guil. Hormani Caesariburgensis*, STC 13811 (London: Printed by Richard Pynson, 1519), fol. 80v; N8v. Nicholas Orme describes the volume as "English sentences with model Latin translations." Nicholas Orme, "Horman, William (1457–1535)," *Oxford Dictionary of National Biography*, accessed June 26, 2017, http://www.oxforddnb.com/view/article/13779.

20. *OED Online*, s.v. “water-leaf, *n.*,” December 2018, Oxford University Press, www.oed.com.

21. Joseph Jérôme Le Français de Lalande, *Art de faire le papier. Description des arts et métiers* (Paris: Chez Desaint & Saillant, Libraries [. . .], 1761); Jean-Baptiste le Rond d’Alembert and Denis Diderot, eds., *Encyclopédie, ou Dictionnaire raisonné des sciences, des arts et des métiers*, vol. 5 (Paris: Briasson et al., 1767), plates 10–12. I rely on images from d’Alembert and Diderot primarily because details are clearer in these images than in Lalande. For Lalande’s images, see “Joseph Jerome de la Lande, *Art de Faire le Papier* (Paris, France, 1761),” Cary Graphic Arts Collection Image Database, accessed January 16, 2017, http://library.rit.edu/cary/joseph-jerome-de-la-lande-art-de-faire-le-papier.

22. On the development of rosin-alum internal sizing, see Cathleen Baker, “Nineteenth-Century Papers: Sizing Does Matter,” *Hand Papermaking* 30, no. 2 (Winter 2015): 19–21.

23. Lalande’s illustrations show four workers in each space.

24. Cited in J. H. Thomas, “The Company of White Paper Makers in Hampshire: An Inventory of Plant,” *Post-Medieval Archaeology* 11, no. 1 (January 1977): 32.

25. Note that while it is possible to return the pulp from a damaged sheet back into the vat, a careful papermaker would likely rehydrate—and may even re-beat—such pulp before adding it back in, increasing the labor-time lost.

26. Dard Hunter, *Papermaking: The History and Technique of an Ancient Craft* (New York: Dover, 1943), 194.

27. Hunter, *Papermaking*, 194.

28. Hunter’s use of the word “bruised” to describe less-than-perfect sheets also recalls paper’s animal counterpart. A fascinating example of bruised parchment can be found in Plimpton MS263 at Columbia University’s Rare Books and Manuscript Library. The manuscript served as Wynkyn de Worde’s printer’s copy for *De Proprietatibus Rerum* (1495). Robert W. Mitchner claims there is a “smudge” on fol. 186r of the MS “which looks suspiciously like a printer’s thumb-print.” Robert W. Mitchner, “Wynkyn de Worde’s Use of the Plimpton Manuscript of *De Proprietatibus Rerum*,” *The Library* 6, no. 1 (June 1951): 9. Closer inspection, with the generous assistance of Conseulo Dutschke and Alexis Hagadorn at Columbia, revealed that the dark spots on the MS are actually the result of blood that was trapped in the skin at the time of slaughter, likely the result of a bruise. Of particular interest to me is the mistaken identity: what was thought to be the physical imprint of a significant printer turns out to be a beastly physical defect. On bruises and MS stains due to blood trapped in the skin, see Christopher Clarkson, “Rediscovering Parchment: The Nature of the Beast,” *Paper Conservator* 16 (1992): 5–26.

29. Baker, “Nineteenth-Century Papers,” 19.

30. One promising avenue I cannot pursue in this chapter is the regular social/business interchange between papermakers, on the one hand and butchers, tanners, etc., on the other hand. Ian Douglas, focusing on a later period, but one in which bones would have been used for sizing, estimates that in nineteenth-century cities “bones formed 20 per cent of animal wastes and had many uses.” Ian Douglas, *Cities: An Environmental History* (New York: I.B. Tauris, 2013), 58. Simon Barcham Green claims that Hayle Mill may well have relied on a local tanning operation, which “produced a variety of by-products including unusable parts of the hide which contain collagen.” By the early twentieth century, however, “for economics and reliability, Hayle Mill began to buy in gelatin as a granular material from large factories elsewhere.” Green, “Sizing at Hayle Mill,” 11–12.

31. John Carter and Nicolas Barker, *ABC for Book Collectors*, 8th ed. (New Castle, DE: Oak Knoll Press, 2004), 206.

32. Philip Gaskell, *A New Introduction to Bibliography* (New Castle, DE: Oak Knoll Press, 2009), 59.

33. John Florio, *A Worlde of Wordes*, STC 11098 (London: By Arnold Hatfield for Edw. Blount, 1598), 61; F1r. Florio’s dictionary also has entries for “Sisa, *a kind of syse or glew that painters vse*” and “Sisare, *to syse, to glew, to stiffen*,” 373; Ii1r.

34. Hunter, *Papermaking*, 62. Hunter does recognize that sizing could also be made from “parings of hides” (194).

35. Imberdis, *Papyrus*, 57.

36. Hunter, *Papermaking*, 194.

37. Józef Dabrowski and John S. G. Simmons, “Permanence of Early European Hand-Made Papers,” *Fibers and Textiles in Eastern Europe* 11, no. 1 (March 2003): 9. The quotation in the original appears on 115v; the English translation, from Latin, is from Dabrowski and Simmons. On the use of parchment

scraps for sizing, see also Karen Garlick, "A Brief Review of the History of Sizing and Resizing Practices," *Book and Paper Group Annual* 5 (1986), cool.conservation-us.org.

38. John Bidwell, "French Paper in English Books," in *The Cambridge History of the Book in Britain*, vol. 4, *1557–1695*, ed. John Barnard and D. F. McKenzie (Cambridge: Cambridge University Press, 2002), 586.

39. The rag-and-bone man turns up in popular culture in places as varied as Laura Ingalls Wilder's *Farmer Boy*, early Betty Boop cartoons, and Bob Dylan's "Stuck Inside of Mobile with the Memphis Blues Again." Rag-and-bone collectors remain active internationally through at least the 1950s. In the later years, the value of bone collecting had to do with fertilizer more than with gelatin or glue. Laura Ingalls Wilder, *Farmer Boy* (New York: HarperTrophy, 2007); Bob Dylan, "Stuck Inside of Mobile with the Memphis Blues Again," track 6 on *Blonde on Blonde*, Columbia Records, 1966; Dave Fleischer, *Any Rags?* (New York: Paramount Pictures, 1932). On the use of phosphorus-rich bones in fertilizer, see also B. W. Clapp, *An Environmental History of Britain Since the Industrial Revolution* (London: Routledge, 2014), 229–30.

40. Bidwell, "French Paper in English Books," 586.

41. Joseph de la Lande, *The Art of Making Paper* (Leicestershire: Plough Press, 1978), 23.

42. Timothy Barrett et al., "Project Goals," Paper Through Time: Nondestructive Analysis of 14th- Through 19th-Century Papers, The University of Iowa, last modified November 28, 2011, http://paper.lib.uiowa.edu /goals.php.

43. Barrett et al., Paper Through Time.

44. Barrett et al., "Conclusions," Paper Through Time, last modified July 1, 2013, http://paper.lib.uiowa.edu/conclusions.php.

45. Barrett et al., "Discussion," "Conclusions," "Results."

46. Barrett et al., "Non-Chronological Plots," Paper Through Time, last modified January 18, 2012, http://paper.lib.uiowa.edu/nonchron.php; Cathleen A. Baker, *From the Hand to the Machine: Nineteenth-Century American Paper and Mediums; Technologies, Materials, and Conservation* (Ann Arbor, MI: The Legacy Press, 2010).

47. Barrett, "European Papermaking Techniques, 1300–1800." See also Simon Barcham Green, who claims, based on his experience at Hayle Mill, that "cooling the sized sheets . . . helped to lower the incidence of under-sizing or uneven sizing," but also notes that, "despite best practices, gelatin sizing typically has a very high failure rate." Barcham Green, "Sizing at Hayle Mill," 12, 17n6.

48. James Mosley, "The Technologies of Print," in *The Book: A Global History*, ed. Michael F. Suarez and H. R. Woudhuysen (Oxford: Oxford University Press, 2013), 133.

49. Lotte Hellinga, "Printing," in *The Cambridge History of the Book in Britain*, vol. 3, *1400–1557*, ed. Lotte Hellinga and J. B. Trapp (Cambridge: Cambridge University Press, 1999), 92. There is some disagreement among scholars and artisans as to whether unsized paper should be dampened before printing. However, even if one grants that a printer would have dampened all paper—sized and unsized—the sized paper ends up causing more labor for at least two reasons: (1) being less absorbent it requires more water to dampen; (2) as Cathleen Baker observes, "Mildewing could occur in packs of damp paper run through the press once and then stored covered so that the sheets would not dry out until the second side could be printed," and sized paper was much more likely to mildew in these conditions. Baker, "Nineteenth-Century Papers," 20.

50. Ronald B. McKerrow, *An Introduction to Bibliography for Literary Students* (Oxford: Clarendon Press, 1928), 116.

51. Hunter, *Papermaking*, 194. Unfortunately, Hunter does not identify his data sample.

52. Theodore L. De Vinne, *The Invention of Printing: A Collection of Facts and Opinions* (New York: F. Hart, 1878), 537–38.

53. J. A. Szirmai, *The Archaeology of Medieval Bookbinding* (Brookfield, VT: Ashgate, 1999), 177.

54. Bernard C. Middleton, "Gluing the Spine," in *A History of English Craft Bookbinding Technique* (London: The Holland Press, 1963), 54. Middleton explains that "gluing up" was used in the process of rounding and backing, and "as these two processes were not introduced before the early part of the sixteenth century glue was not used before then. Examination of the backs of sections [of books] suggests that in the second half of the sixteenth century, and even as late as the early part of the seventeenth, a few binders did not 'glue up'" (54). See also Szirmai, *Medieval Bookbinding*, 191.

55. Szirmai attributes this insight to Paul Adam, *Das Restuarieren alter Bücher* (Halle [Saale]: Verlag Wilhelm Knapp, 1927). Szirmai, *Medieval Bookbinding*, 191.

56. Nicholas Pickwoad, "Onward and Downward: How Binders Coped with the Printing Press Before 1800," in *A Millennium of the Book: Production, Design, and Illustration in Manuscript and Print, 900–1900*, ed. Robin Myers and Michael Harris (New Castle, DE: Oak Knoll Press, 1994), 63.

57. Pickwoad, "Onward and Downward," 63.

58. See "The Book-binder," in Johannes Amos Comenius, *Orbis Sensualium Pictus*, trans. Charles Hoole, Wing C5523 (London: Printed for J. Kirton, 1659), 194–95; O1v-2r.

59. See Comenius, *Orbis Sensualium Pictus*, 188–89; N6v-7r.

60. Comenius, 189; N7r.

61. In the 1693 poem cited earlier, which was composed in Latin, Imberdis uses "glutinis" and "glutine charta" to refer to a solution that is made from animal parts. Imberdis, *Papyrus*, 56, 60.

62. N. Katherine Hayles, "The Complexities of Seriation," Mobile Citizens, Media States," special issue, *PMLA* 117, no. 1 (January 2002): 119. For a fuller discussion of seriation and skeuomorphs, see N. Katherine Hayles, *How We Became Posthuman: Virtual Bodies in Cybernetics, Literature, and Informatics* (Chicago: University of Chicago Press, 1999), especially 13–18.

63. See Middleton, *Craft Bookbinding*, 127–29. According to Middleton, clasps were "an important feature of medieval books" (127), an ornamental feature of sixteenth-century books (128), and a "pretentious" feature of Victorian books (129).

64. Middleton, *Craft Bookbinding*, 128.

65. Middleton, 128–29.

66. Hayles, "Seriation," 119.

67. Russell Maret, "Hungry Bibliophiles: An Experiment in Utilitarian Bookmaking," *Hand Papermaking* 30, no. 2 (Winter 2015): 30–31.

68. Sir Hugh Plat, *The Iewell House of Art and Nature Conteining Diuers Rare and Profitable Inuentions, Together with Sundry New Experimentes in the Art of Husbandry*, STC 19991 (London: Printed by Peter Short, 1594), 46; H3v. Alan Stewart quotes this passage and adds the helpful insertion "red arsenic sulphide"; Alan Stewart, *Shakespeare's Letters* (Oxford: Oxford University Press, 2008), 44.

69. Helen Smith, "'A unique instance of art': The Proliferating Surfaces of Early Modern Paper," *Journal of the Northern Renaissance* 8 (2017): par. 10, 2.

70. Robert Darnton, "'What Is the History of Books?' Revisited," *Modern Intellectual History* 4, no. 3 (2007): 498.

71. *OED Online*, s.v. "sink, *v.*" December 2016, Oxford University Press, http://www.oed.com.

72. Mathurin Cordier, *Corderius Dialogues Translated Grammatically for the More Speedy Attaining to the Knowledge of the Latine Tongue*, STC 5762 (London: Printed by Humfrey Lownes, for Thomas Man, 1614), 154–55; L5v-L6r. Cordier's 1614 text was reprinted four more times by 1653.

73. Cordier, *Corderius Dialogues*, 245; R3r.

74. John Preston, *The Saints Svbmission, and Sathans overthrow*, STC 20266 (London: Printed by J. D[awson] and are to be sold by Peter Cole at the signe of the Glove in Cornhill, neere the Royal-Exchange, 1638), 177–78; I5r-v. See also *The Holy Bible* [*The King James Bible*], STC 2217 (London: Robert Barker, 1611), V2r.

75. Thomas Fuller, *Davids Hainovs Sinne*, STC 11463 (London: Printed by Tho. Cotes, for Iohn Bellamie, dwelling at the three Golden Lyons in Cornehill, 1631), B4v, stanza 40, line 4.

76. Fuller, *Davids Hainovs Sinne*, B2v, stanzas 31 and 32.

77. *OED Online*, s.v. "† sepian, *adj.*," December 2018, Oxford University Press, http://www.oed.com. Not surprisingly, given the 1631 date of Fuller's work, blotting is central to the plot in the next part of the poetic narrative, titled *Davids hearty Repentance*. There the Lord retrieves his (printed) book of life and "Within this Booke, hee sought for *Davids* name, / Which having found, he proffered to blot"; he is stopped from the action only by the intercession of Jesus, the "the Spotlesse lambe." Fuller, *Davids Hainovs Sinne*, B8r-v, stanzas 5–8, esp. 6.1–2, 8.1.

78. On "spungy" paper, see Joseph Moxon, *Mechanick Exercises: Or, the Doctrine of Handy-Works*, Wing M3013–3014, vol. 2 (London: Printed for Joseph Moxon, 1677), section XXIV, 304; Ss2v. I discuss staining, blotting, and King David's most notable sin in the previous chapter.

79. Barcham Green, "Sizing at Hayle Mill," 14.

80. Richard Harvey, *Plaine Percevall the Peace-Maker of England*, STC 12914 (London: Printed in Broad- streete at the signe of the Pack-staffe [i.e., Eliot's Court Press for G. Seton], 1590), 24; D4v.

81. Henry Cary Falkland, *The Mariage Night Written by the Lord Viscount Fawlkland*, Wing F315 (London: printed by W. G[odbid] for R. Crofts at the Crown in Chancery-Lane under Sergeants- Inne, 1664), 11; C2r.

82. Margaret Spufford, *Small Books and Pleasant Histories: Popular Fiction and Its Readership in Seventeenth-Century England* (Cambridge: Cambridge University Press, 1981), 48; Heidi Brayman, *Reading Material in Early Modern England: Print, Gender, and Literacy* (Cambridge: Cambridge University Press, 2005), 81.

83. William Cornwallis, *Essayes*, STC 5775 (London: Printed [by S. Stafford and R. Read] for Edmund Mattes, 1600–1601), essay 15 ("Of the Observation, and Use of Things"), I7r.

84. Thomas Nashe, *The Vnfortunate Traueller. Or, The life of Iacke Wilton Newly corrected and augmented*, STC 18381 (London: Imprinted by Thomas Scarlet for Cuthbert Burby, 1594), A2r. Of this preface, Julian Yates writes, "The pun on 'privy,' meaning both private and toilet, stages this 'precious' intimacy in the contact of the pages with the body of the reader as toilet paper." Julian Yates, *Error, Misuse, Failure: Object Lessons from the English Renaissance* (Minneapolis: University of Minnesota Press, 2003), 107.

85. According to Blair, "The increased availability of paper coincided with a rise in paper records of all kinds, from diplomatic correspondence to family papers to collections of scholarly notes." Ann Blair, "The Rise of Note-Taking in Early Modern Europe," *Intellectual History Review* 20, no. 3 (2010): 310.

86. Blair, "The Rise of Note-Taking," 307. According to Pettegree, "That the wisdom of the ancients survived at all we owe to two major changes in the third and fourth centuries: the replacement of papyrus with parchment, and of the scroll with the codex." Pettegree, *The Book in the Renaissance*, 5.

87. See "Seventeenth-Century Letters Discovered at Knole," National Trust, accessed July 7, 2017, https://www.nationaltrust.org.uk/knole/features/seventeenth-century-letters-discovered-at-knole.

88. Baker, "Nineteenth-Century Papers," 21n1.

89. Timothy Barrett and Cynthea Mosier, "The Role of Gelatin in Paper Permanence," *Journal of the American Institute for Conservation* 34, no. 3 (Autumn–Winter 1995): 184.

90. Barrett and Mosier, "The Role of Gelatin," 184.

91. Barrett et al., "Conclusions," http://paper.lib.uiowa.edu/conclusions.php#. For a study of contemporary sizing methods that "play a major role in promoting or inhibiting fungal growth," see Flavia Pinzari et al., "Biodeterioration of Paper: A SEM Study of Fungal Spoilage Reproduced Under Controlled Conditions," *Macromolecules in Cultural Heritage* 238, no. 1 (April 2006): 64.

92. John Baty, "Sizing Science: Can Physical Science and Engineering Answer Some of Our Lingering Questions?," *Hand Papermaking* 30, no. 2 (Winter 2015): 10.

93. B. L. Browning, "The Nature of Paper," *Library Quarterly: Information, Community, Policy* 40, no. 1 (January 1970): 31.

94. Alan B. Farmer, "Playbooks and the Question of Ephemerality," in *The Book in History, the Book as History: New Intersections of the Material Text*, ed. Heidi Brayman, Jesse M. Lander, and Zachary Lesser (New Haven: Beinecke Rare Book & Manuscript Library, Yale University, 2016), 97.

95. Farmer, "Playbooks," 116.

96. Joseph A. Dane, *Abstractions of Evidence in the Study of Manuscripts and Early Printed Books* (Farnham: Ashgate, 2009), 19.

97. Malcolm Walsby, *The Printed Book in Brittany, 1484–1600* (Leiden: Brill, 2011), 99.

98. Flavia Bruni and Andrew Pettegree, eds., *Lost Books: Reconstructing the Print World of Pre-Industrial Europe* (Leiden: Brill, 2016). See book description on the publisher's website "About," *Lost Books: Reconstructing the Print World of Pre-Industrial Europe*, Brill, accessed July 2017, https://brill.com/abstract/title/32814. See also Jonathan Green, Frank McIntyre, and Paul Needham, "The Shape of Incunable Survival and Statistical Estimation of Lost Editions," *Papers of the Bibliographic Society of America (PBSA)* 105, no. 2 (June 2011): 141–75; and Goran Proot and Leo Egghe, "Estimating Editions on the Basis of Survivals: Printed Programmes of Jesuit Plays in *Provincia Flandro-Belgica* before 1773, with a Note on the 'Book Historical Law,'" *PBSA* 102, no. 2 (June 2008): 149–74.

99. Alexandra Hill, "Lost Print in England: Entries in the Stationers' Company Register, 1557–1640," in Bruni and Pettegree, *Lost Books*, 159.

CHAPTER 5

Note to epigraph: Wendy Cope, *Making Cocoa For Kingsley Amis* (London: Faber and Faber, 1986), 59.

1. Jeffrey Todd Knight, *Bound to Read: Compilations, Collections, and the Making of Renaissance Literature* (Philadelphia: University of Pennsylvania Press, 2013), 180.

2. The value of detailed, descriptive item notes is not in question, but rather an unrealistic standard of perfection that suggests well-used books are "imperfect." Examples abound and can be found most easily by searching "imperfect" in online catalogues. For examples, a copy of the Fourth Folio at the Folger Shakespeare Library is "tattered and torn with a burned spot in some leaves, affecting text . . . and many page numbers. . . . Other imperfections affect text"; a legal volume at the Huntington Library is "IMPERFECT: leaf 89 ripped at bottom, with no loss of text"; and Donne's *Poems* (1650) at the Harry Ransom Center is "imperfect: wanting the initial blank leaf. . . . effaced inscription on recto of frontispiece portrait; leaf 2C1 torn, mended." Catalogue entry for Folger call number S2915 Fo.4 no.18, Shakespeare et al., *Mr. William Shakespear's Comedies, Histories, and Tragedies: Published According to the True Original Copies: Unto Which Is Added, Seven Plays, Never before Printed in Folio*, 4th ed., Wing S2915 (London: Printed for H. Herringman, 1685), *Hamnet: Folger Library Catalogue*, accessed December 10, 2018, http://hamnet.folger.edu/cgi-bin/Pwebrecon.cgi?BBID=108071; catalogue entry for Huntington call number 497071, Edward Coke et al., *Le tierce part des Reportes del Edward Coke l'attorney generall le Roigne* [. . .], STC 5499.2 ([London: Printed by Adam Islip] in aedibus Thomae Wight, 1602), *Huntington Library Catalogue*, accessed December 10, 2018, https://catalog.huntington.org/record=b1066975; catalogue entry for University of Texas at Austin Library call number PFORZ Add. 22, John Donne, John Marriot, and Richard Marriot, *Poems*, Wing D1869 (London: Printed for John Marriot [. . .] and are to be sold by Richard Marriot, 1650), *University of Texas Libraries, Library Catalog*, http://catalog.lib.utexas.edu/record=b3444254~S18.

3. Siân Echard, "House Arrest: Modern Archives, Medieval Manuscripts," *Journal of Medieval and Early Modern Studies* 30, no. 2 (Spring 2000): 203, 186. McGann's "local habitation and a name" quotes Theseus from Shakespeare's *A Midsummer Night's Dream:*

> The poet's eye, in a fine frenzy rolling,
> Doth glance from heaven to Earth, from Earth to heaven,
> And as imagination bodies forth
> The forms of things unknown, the poet's pen
> Turns them to shapes and gives to airy nothing
> A local habitation and a name.
> (5.1.12–18)

4. Jennifer Summit, *Memory's Library: Medieval Books in Early Modern England* (Chicago: University of Chicago Press, 2008), 2.

5. Summit, *Memory's Library*, 2. See also Bruce Holsinger, "Of Pigs and Parchment: Medieval Studies and the Coming of the Animal," *PMLA* 124, no. 2 (March 2009): 616–23.

6. See also Jennifer Borland, "Unruly Reading: The Consuming Role of Touch in the Experience of a Medieval Manuscript," in *Scraped, Stroked, and Bound: Materially Engaged Readings of Medieval Manuscripts*, ed. Jonathan Wilcox (Turnhout: Brepols, 2013), 97–114. Borland notes that "when we touch the book today, we not only connect to the book's moment of creation, but also are reminded of the 800 years of readers, viewers, and handlers" (114).

7. See Matthew 6:19: "Lay not vp for yourselues treasures vpon earth, where moth and rust doth corrupt, and where thieues breake thorow and steale." *The Holy Bible* [*The King James Bible*], STC 2217 (London: By Robert Barker, 1611).

8. Simon C. Estok, "Afterword: Ecocriticism on the Lip of a Lion," in *Ecocritical Shakespeare*, ed. Lynne Bruckner and Dan Brayton (Farnham: Ashgate, 2011), 242.

9. According to Estok, "much of the research" for a monograph he published the same year as the essay cited above "was carried out in the Folger Shakespeare Library." Simon C. Estok, *Ecocriticism and Shakespeare: Reading Ecophobia* (New York: Palgrave Macmillan, 2011), ix.

10. "Experts' Roundtable on Sustainable Climate Management Strategies" (roundtable, Getty Conservation Institute: Alternative Climate Controls for Historic Buildings, Tenerife, Spain, April 2007), 1. The 98-page edited transcript is available via a dedicated page on GCI's website that also includes twelve essays by participants and an "Experts' Bibliography." Participants include Giacomo Chiari, Dario Camuffo, Ernest Conrad, Kathleen Dardes, Michael Henry, Richard Kerschner, Roman Kozlowski, Shin Maekawa, Stefan Michalski, Tadj Oreszczyn, Tim Padfield, Henk Schellen, Sarah Staniforth, and Franciza Toledo. See http://www.getty.edu/conservation/our_projects/science/climate/roundtable_transcript.pdf.

11. "Experts' Roundtable," 1.

12. Ernest A. Conrad, "Climate Control Systems Design and Climate Change," (discussion paper, Getty Conservation Institute: Alternative Climate Controls for Historic Buildings, Tenerife, Spain, April 2007), 1. Quotations are from Conrad's submitted discussion paper as provided on GCI's website, http://www.getty.edu/conservation/our_projects/science/climate/paper_conrad.pdf.

13. Conrad, "Climate Control Systems," 2 and 5.

14. Sarah Staniforth, "Slow Conservation," *Studies in Conservation* 55, no. 2 (2010): 76. In 2016, after completing a successful "Good Energy" pilot program begun in 2013, the National Trust committed "£30 million in renewable energy . . . to reduce our reliance on fossil fuels, cut energy usage by 20 per cent and source 50 per cent of our energy from renewable sources on land we look after by 2020."See "£30 Million Investment in a Sustainable Future," National Trust, August 8, 2016,https://www.nationaltrust.org.uk/news/30-million-investment-in-a-sustainable-future.

15. John Donne, "A Valediction: Of the Book," in *John Donne: The Complete English Poems*, ed. A. J. Smith (London: Penguin, 1974), 85–87. All quotations from Donne's poetry, unless otherwise noted, are from this edition.

16. On Renaissance expectations for book survival, see Adam Smyth, "Burning to Read: Ben Jonson's Library Fire of 1623," in *Book Destruction from the Medieval to the Contemporary*, ed. Adam Smyth and Gill Partington (New York: Palgrave Macmillan, 2014), esp. the opening section, "Dying Books," 34–37.

17. On the impending threat of a global, mass extinction referred to as the sixth extinction, see especially Elizabeth Kolbert, *The Sixth Extinction: An Unnatural History* (New York: Henry Holt, 2014).

18. *OED Online*, s.v. "biodeterioration, *n.*," July 2018, Oxford University Press, www.oed.com; H. J. Hueck, "The Biodeterioration of Materials as Part of Hylobiology," *Material und Organismen* 1 (1965): 7. See also Dennis Allsopp, Kenneth Seal, and Christine Gaylarde, *Introduction to Biodeterioration*, 2nd ed. (Cambridge: Cambridge University Press, 2004), 1; and Giulia Caneva, Maria Pia Nugari, and Ornella Salvadori, eds., *Plant Biology for Cultural Heritage: Biodeterioration and Conservation*, trans. Helen Glanville (Los Angeles: Getty Conservation Institute, 2008), 15.

19. Jonathan Bate, *The Song of the Earth* (Cambridge, MA: Harvard University Press, 2000), 92.

20. Henry Vaughan, "The Book," in *Silex Scintillans: Sacred Poems and Private Ejaculations,* 2nd ed., Wing V126 (London: Printed for Henry Crips and Lodowick Lloyd, 1655), 2nd book, 90–81 [*sic*]; G8v-H1r.

21. Edmund Spenser, *The Faerie Queene*, ed. A. C. Hamilton with Hiroshi Yamashita, Toshiyuki Suzuki, and Shohachi Fukuda, 2nd ed. (London: Routledge, 2001), bk. 2, canto 11; 261–69.

22. On how the poem ironically and "mimetically reproduces the historical effacement of the women's identities" even as it claims that the anonymous beloved will be remembered, see Elizabeth D. Harvey, "Ventriloquizing Sappho: Ovid, Donne, and the Erotics of the Feminine Voice," *Criticism* 31, no. 2 (Spring 1989): 115. Both Piers Brown and Megan Heffernan discuss the leave-taking function of Donne's iconic valediction poems; see, respectively, Piers Brown, "Donne's Hawkings," *Studies in English Literature, 1500–1900* 49, no. 1 (Winter 2009): 67–86; and Megan Heffernan, "Turning Sonnets into Poems: Textual Affect and John Benson's Metaphysical Shakespeare," *Shakespeare Quarterly* 64, no. 1 (Spring 2013): 71–98.

23. Charles Eliot Norton includes just the first three stanzas in his 1905 edition of *The Love Poems of John Donne*, noting the "superb climax" of the first three stanzas and suggesting that the "omitted four stanzas are of little worth"; see John Donne, *The Love Poems of John Donne*, ed. Charles Eliot Norton (Boston: Houghton, Mifflin, 1905), 84. George Williamson, considering the progression of the

stanzas, notes "the increasing cynicism with which he [the poet] has described her book." George Williamson, *Six Metaphysical Poets: A Reader's Guide* (Syracuse: Syracuse University Press, 1967), 66.

24. Peter McCullough, "Donne to Death," The Bodleian Libraries (BODcasts), May 13, 2016, https://podcasts.ox.ac.uk/donne-death, 1:07. McCullough further claims that "all of his preordination prose takes death in some way as its subject. . . . Following his ordination, though, in 1615 death was still never far from Donne's creative mind" (6:35). Emphasizing a similar if slightly variant preoccupation in Donne's work—the other side of the coin, so to speak—Raymond-Jean Frontain claims that "resurrection became the central trope of his imagination, informing his religious as well as his amorous work." Raymond-Jean Frontain, "Donne's Imperfect Resurrection," *Papers on Language & Literature* 26, no. 4 (Fall 1990): 539.

25. Vaughan's poem does momentarily look to the future, but his concern is with "trees, beasts and men": "Give him amongst thy works a place, / Who in them lov'd and sought thy face!" (ll. 29–30).

26. Natural deterioration over time does not necessarily preclude disintegration at the day of resurrection. Donne's poetry is filled with references to decay and to resurrection. I follow Gary Waller in thinking that "Donne's lyrics are notoriously indifferent to consistency and closure. . . . Any meanings which we may confidently take from a particular reading disintegrate as we leap from one poem to the next." Gary Waller, *English Poetry of the Sixteenth Century*, 2nd ed. (New York: Longman, 1993), 226.

27. See Joshua Calhoun, "Ecosystemic Shakespeare: Vegetable Memorabilia in the *Sonnets*," *Shakespeare Studies* 39 (2011): 64–73.

28. Ben Jonson, "To the memory of my beloued, The Author," in *Mr. VVilliam Shakespeares comedies, histories, & tragedies: published according to the true originall copies*, by William Shakespeare, ed. Iohn Heminge and Henry Condell, STC 22273 (London: Printed by Isaac Iaggard, and Ed. Blount, 1623), A4r. See image from the Horace Howard Furness Shakespeare Library of the Schoenberg Center for Electronic Text & Image (SCETI), accessed October 1, 2010, http://sceti.library.upenn.edu/sceti/printed booksNew/index.cfm?Te xtID = firstfolio&PagePosition = 9. I cite a version of the First Folio accessed via the publicly available SCETI database because, paradoxically, electronic databases like SCETI have made the plant and cloth nature of Renaissance books more visible to those who may not have access to them in rare books collections. At 8x magnification, these high-quality images reveal flecks of plant and cloth matter in many of the Folio's pages.

29. Martin E. Weaver, with F. G. Matero, *Conserving Buildings: Guide to Techniques and Materials* (New York: Wiley, 1993), 85. See also Guy Munsch, "Condition Report for the Folger Shakespeare Library, July 2002," 4. I am grateful to Erin Blake, head of Collection Information Services, for generous assistance, for sharing images and documentation, and for calling my attention to key details about the Folger's built environment and its history.

30. Munsch, "Condition Report," 14. Ultimately, Munsch, the architectural conservator, suggested that "the best that conservation will be able to offer for these bas-relief sculptures in their exposed, exterior environment is that we can try to slow down the deterioration process as much as possible" (14).

31. Shobha Shrivastava, "Biodeterioration of Art Objects on Paper and Their Conservation," *Research Journal of Recent Sciences* 4 (2015): 42.

32. Flavia Pinzari, Giovanna Pasquariello, and Antonella De Mico, "Biodeterioration of Paper: A SEM Study of Fungal Spoilage Reproduced Under Controlled Conditions," *Macromolecules in Cultural Heritage* 238, no. 1 (April 2006): 58. As the authors explain, "Microbial activity affects not only the appearance of the objects . . . but often consists of a deep modification of their chemical and physical structure" (58).

33. Alan Farmer estimates the "overall loss rate for the entire book trade (excluding single-leaf publications) from 1580 to 1640" to be 23.3 percent. Alan Farmer, "Playbooks and the Question of Ephemerality," in *The Book in History, the Book as History: New Intersections of the Material Text*, ed. Heidi Brayman, Jesse M. Lander, and Zachary Lesser (New Haven: Beinecke Rare Book & Manuscript Library, Yale University, 2016), 106. It is worth bearing in mind that loss rates are a topic of much scholarly debate and, as I have suggested in Chapter 3, they do not account for the crucial presence or absence of gelatin sizing.

34. Andrew Pettegree, "The Legion of the Lost: Recovering the Lost Books of Early Modern Europe," in *Lost Books: Reconstructing the Print World of Pre-Industrial Europe*, ed. Andrew Pettegree and Flavia Bruni (Leiden: Brill, 2016), 9.

35. Alan Galey, *The Shakespearean Archive: Experiments in New Media from the Renaissance to Postmodernity* (Cambridge: Cambridge University Press, 2014), 49; see also 46–56 for Galey's full reading of the National Archives Building.

36. Francis Bacon, *The New Organon*, ed. Lisa Jardine, trans. Michael Silverthorne (Cambridge: Cambridge University Press, 2000), 228.

37. Bacon, *The New Organon*, 211.

38. Bacon, 211, 212.

39. James Spedding, ed., *The Works of Francis Bacon*, vol. 14 (London: Longmans, Green, Reader, and Dyer, 1874), 549–50. See also Markku Peltonen, "Bacon, Francis, Viscount St Alban (1561–1626), lord chancellor, politician, and philosopher," *Oxford Dictionary of National Biography*, accessed August 26, 2016, http://www.oxforddnb.com/view/article/990.

40. Spedding, *Works of Francis Bacon*, 14:550.

41. On "archival temples," see Eric Ketelaar, "Archival Temples, Archival Prisons: Modes of Power and Protection," *Archival Science* 2, no. 3 (2002): 221–38.

42. For a particularly famous example of a Shakespearean text that survived for two hundred years in a closet, see Zachary Lesser's discussion of Q1 *Hamlet* in *"Hamlet" After Q1: An Uncanny History of the Shakespearean Text* (Philadelphia: University of Pennsylvania Press, 2015).

43. As Adam Smyth has argued, although "Ben Jonson's conspicuous attempts to fix and authorize his texts have been taken as paradigmatic," our scholarly preoccupation with "textual fixity" often overlooks "the importance of textual transmission and malleability." Adam Smyth, "'Art Reflexive': The Poetry, Sermons, and Drama of William Strode (1601?–1645)," *Studies in Psychology* 103, no. 4 (Fall 2006): 456, 463.

44. Raymond Arsenault, "The End of the Long Hot Summer: The Air Conditioner and Southern Culture," *Journal of Southern History* 50, no. 4 (November 1984): 601.

45. Gail Cooper, *Air-Conditioning America: Engineers and the Controlled Environment, 1900–1960* (Baltimore: Johns Hopkins University Press, 2002), 24.

46. Willis H. Carrier, "The Economics of Man-Made Weather," *Scientific American* 148, no. 4 (April 1933): 200.

47. Emily Badger and Alan Blinder, "How Air-Conditioning Conquered America (Even the Pacific Northwest)," *New York Times*, August 4, 2017.

48. Stefan Michalski, writing in 2007, claims that "until recently, climate specifications for museums and archives in the American Society of Heating, Refrigerating, and Air-Conditioning Engineers *ASHRAE Handbook* (the North American HVAC engineer's bible) was buried in a chapter on miscellaneous special applications. The advice was simple: provide 21C and 50% RH, with minimal fluctuations." Stefan Michalski, "The Ideal Climate, Risk Management, the ASHRAE Chapter, Proofed Fluctuations, and Towards a Full Risk Analysis Model" (discussion paper, Getty Conservation Institute: Alternative Climate Controls for Historic Buildings, Tenerife, Spain, April 2007), 2. Quotations are from Michalski's submitted discussion paper as provided on GCI's website, http://www.getty.edu/conservation/our_projects/science/climate/paper_michalski.pdf. Sarah Staniforth writes that "one of the drivers for maintaining constant temperature and RH has been compliance with rigid standards introduced in the 1980s for international loan exhibitions." Staniforth, "Slow Conservation," 78; note that Staniforth also cites ongoing conversations and interim guidelines among institutions for "reconciling the desirability of the long-term preservation of collections with the need to reduce energy cost" (79).

49. One example of how a book might move from place to place while being kept in the same or similar conditions in each place is the recent "First Folio Tour" in which the Folger Shakespeare Library sent a 1623 Folio to each of the fifty states, Washington, DC, and Puerto Rico. See, for example, Emily Litvack, "First Folio Fights Off Agents of Decay," Research, Discovery, & Innovation, The University of Arizona, March 7, 2016, https://research.arizona.edu/stories/first-folio-fights-agents-decay. See also "About the First Folio Tour," Folger Shakespeare Library, accessed December 10, 2018, http://www.folger.edu/about-the-first-folio-tour.

50. See "Major Biomes of the World," World Wildlife Fund, accessed December 10, 2018, http://wwf.panda.org/about_our_earth/teacher_resources/webfieldtrips/major_biomes. As the WWF notes, "Scientists argue on the exact number, or different types of biomes in existence"; scientists argue, too, over the precise definition of "biome" and what should and should not count as one. WWF links to

a map of world biomes for further information on subcategories. See http://www.worldbiomes.com/biomes_map.htm.

51. Each biome classification has a number of subcategories.

52. Erle C. Ellis and Navin Ramankutty, "Putting People in the Map: Anthropogenic Biomes of the World," *Frontiers in Ecology and the Environment* 6, no. 8 (October 2008): 439, 445.

53. On growing interest in the field, Brent Stephens writes, "Over the last decade, there has been a dramatic increase in the use of high-throughput molecular techniques to analyze microbial communities in indoor environments"; Brent Stephens, "What Have We Learned About the Microbiomes of Indoor Environments?," *mSystems* 1, no. 4 (July–August 2016): 1.

54. See "About the Study—Microbiomes of the Built Environment," The National Academies of Sciences, Engineering, and Medicine, accessed August 18, 2017, http://nas-sites.org/builtmicrobiome/about-the-study-2. The National Academies of Sciences, Engineering, and Medicine is concluding a twenty-month consensus study on "microbiomes of the built environment." This study is sponsored by the Alfred P. Sloan Foundation, U.S. Environmental Protection Agency (EPA), National Aeronautics and Space Administration (NASA), and National Institutes of Health (NIH).

55. See Stephens, "What Have We Learned?," 2–4.

56. Alexander Mahnert, Christine Moissl-Eichinger, and Gabriele Berg, "Microbiome Interplay: Plants Alter Microbial Abundance and Diversity Within the Built Environment," *Frontiers in Microbiology* 6 (2015): 10.

57. It is worth pointing out that the Folger, far from doing it the way it has always been done, has been actively exploring ways to "create a preservation environment as good or better while consuming less energy." See Stephen C. Enniss and Erin C. Blake, "Folger Sustainable Preservation Environment Project White Paper," Folger Shakespeare Library, November 30, 2011, 2. As part of a NEH Sustaining Cultural Heritage planning grant, they recently completed phase 1 of the "Folger Sustainable Preservation Environment Project," which has revealed ways to cut energy expenses by 46 percent while also improving preservation conditions for the rare materials. The full report is available at https://hcommons.org/deposits/item/hc:12487.

58. Caneva, Nugari, and Salvadori, *Plant Biology for Cultural Heritage*, 173.

59. Caneva, Nugari, and Salvadori, 173.

60. Smyth, "Burning to Read," 43–44.

61. Smyth, 34. See also Smyth's suggestion that defiant lines about outliving death and decay in Shakespeare's *Sonnets* "might thus be read not as descriptions of bibliographical norms, but rather as counter-voices whose wit and urgency derived from their inversion of a prevailing culture of textual transience" (35).

62. Michalski, "The Ideal Climate," 17.

63. Michalski, 17.

64. *OED Online*, s.v. "fossil fuel, *n.*," June 2016, Oxford University Press, http://www.oed.com; Michalski, "The Ideal Climate," 2.

65. Pinzari et al., "Biodeterioration of Paper," 58.

66. MS L.b.541 can be accessed at luna.folger.edu/luna/servlet/s/08t81w. Viewing a digital image of the manuscript on a mineral-rich computer, phone, or tablet screen, we turn the poem back into something closer to stone media format.

67. The page is 31.5 cm in length, but varies in width from 22.2 (bottom) to 22.8 cm (top). "Annæ" appears on a portion of the page that is 22.8 cm and runs from 7.9 to 9.2 cm.

68. These as-yet unidentified flecks are each approximately 1 mm long. Of special interest is a 3 mm long dark fuzz or hair of some kind on the left margin of the recto. A shadowy blur is visible in the high-quality digital image that the Folger has made freely available via Luna, but the contours of the hair are impossible to see in a digital image, even at maximum magnification.

69. Parts of Miriam Jacobson's scholarship on Donne and hair, which I first heard presented as "Vibrant Hair: Temporality and the Undead in Early Modern Poetry" at the Folger Shakespeare Library in summer 2014 and which prompted me to begin actively noticing hair remnants in paper, are included in her as-yet unpublished manuscript, "Renaissance Undead: Reanimating the Past in Early Modern England." I'm grateful to Jacobson for sharing ideas, sources, and conversation.

70. Donne, *The Complete English Poems*, 59.

71. Donne, 75–76.

72. In 2005, the authors of *John Donne's Marriage Letters in the Folger Shakespeare Library* report that although the copy "has generally been accepted as a holograph . . . Mark Bland in *The Manuscripts of Ben Jonson and his Contemporaries* . . . proposes that it is not in Donne's hand, pointing to several anomalous features varying from artifacts known to be in Donne's." M. Thomas Hester, Robert Parker Sorlien, and Dennis Flynn, eds., *John Donne's Marriage Letters in The Folger Shakespeare Library* (Washington, DC: Folger Shakespeare Library; Distributed by University of Washington Press, 2005), 32n55.

73. I cite W. Milgate's translation in John Donne, *The Epithalamions, Anniversaries and Epicedes*, ed. W. Milgate (Oxford: Oxford University Press, 1978), 215.

74. Stow included the epitaph in *The Remains or Remnants of Diverse Worthy Things*, which was added to his 1633 *Survey*. In the column next to the record of Ann Donne's epitaph, Stow lauds the care that was being given to the chapel at the time of writing, noting that "the least defect" is "instantly" repaired. John Stow, *The Survey of London: Contayning the Originall, Increase, Moderne Estate, and Government of that City; Methodically Set downe*, STC 23345 (London: Printed by Elizabeth Pvrslovv, and to be sold by Nicholas Bovrne, 1633), 889; Ffff5r.

75. Hester et al., *John Donne's Marriage Letters*, 32n55.

76. Milgate, ed., Donne, *The Epithalamions*, 78, 215.

77. As if that were not enough mixed mediation, as Milgate notes, these lines recall "Epitaph on Himself," which either is or is imagined as a paper letter (a "funeral scroll") to the Countess of Bedford. Shawcross conjectures that the poem may have been "written in 1608 when Donne was seriously ill." See Milgate, ed., Donne, *The Epithalamions*, 215, 218; John T. Shawcross, ed., *The Complete Poetry of John Donne* (New York: New York University Press, 1968), 248n. The letter becomes a way for Donne to be heard after death. It is a way "when we are speechless grown, to make stones speak." Donne, *The Complete English Poems*, 233; ll. 8, 21.

78. The two manuscripts are Cambridge University Library, Add. ms. 8468 (Narcissus Luttrell ms.) and Harvard Library, ms. Eng. 966.5. For the quotation, see Gary A. Stringer, ed., *The Variorum Edition of the Poetry of John Donne*, vol. 8, *The Epigrams, Epithalamions, Epitaphs, Inscriptions, and Miscellaneous Poems* (Bloomington: Indiana University Press, 1995), 188. The other three extant copies are all at the British Library: Harley 3910, Lansdowne 878, and Lansdowne 948.

79. Ann Rosalind Jones and Peter Stallybrass, *Renaissance Clothing and the Materials of Memory* (Cambridge: Cambridge University Press, 2000), 252.

80. "Explore the Cathedral: South Quire Aisle," St. Paul's Cathedral, accessed September 25, 2016, https://www.stpauls.co.uk/history-collections/history/explore-the-cathedral.

81. John Stubbs, *John Donne: The Reformed Soul: A Biography* (New York: W. W. Norton, 2008), 472.

82. Stubbs, *John Donne*, 473.

83. John Donne, *Deaths dvell, or, A Consolation to the Soule, against the dying Life, and liuing Death of the Body. Deliuered in a Sermon at White Hall, before the Kings Maiesty, in the beginning of Lent, 1630. By that late learned and Reuerend Diuine, Iohn Donne, Dr. in Diuinity, & Deane of S. Pauls, London. Being his last Sermon, and called by his Maiesties houshold The Doctors owne Funerall Sermon*, STC 7031. (London: Printed by Thomas Harper, for Richard Redmer and Beniamin Fisher, and are to be sold at the signe of the Talbot in Alders-gate street, M.DC.XXXII. [1632]). For his 1633 *Poems*, Marriot borrows two verses from *Death's Duel*, Henry King's elegy, and Edward Hyde's epitaph for Donne.

84. [John Donne], *Poems, by J. D. with Elegies on the Authors Death*, STC 7045 (London: Printed by M. F. for Iohn Marriot, and are to be sold at his shop in St *Dunstans* Church-yard in *Fleet-street*, 1633), A2v.

85. Sarah Wall-Randell, "Marlowe, Shakespeare, and the Dead in Sheets," (presentation, "Posthumous Shakespeare" panel, Shakespeare Association of America Conference, New Orleans, March 2016), 4. I am grateful to Wall-Randell for sharing her talk with me in advance of publication. See also Wall-Randell's now-published version of this work: "Marlowe's Lucan: Winding Sheets and Scattered Leaves," in *Christopher Marlowe, Theatrical Commerce, and the Book Trade*, ed. Kirk Melnikoff and Roslyn L. Knutson (Cambridge: Cambridge University Press, 2018) 11–25.

86. I owe this insight to Sarah Wall-Randell, who drew my attention to the uniqueness of Marriot's poem in this regard.

87. Stubbs, *John Donne*, 150.

88. Lara M Crowley, "Archival Research," in *The Oxford Handbook of John Donne (Oxford Handbooks Online)*, ed. Dennis Flynn, M. Thomas Hester, and Jeanne Shami, Oxford Handbooks Online, September 2012, www.oxfordhandbooks.com.

89. M. Thomas Hester, "'Miserrimum dictu': Donne's Epitaph for His Wife," *Journal of English and Germanic Philology* 94, no. 4 (October 1995): 514.

REMAINDERS

1. Dante Alighieri, *Paradiso*, in *The Divine Comedy*, ed. and trans. Charles S. Singleton, vol. 3, pt. 1 (Princeton: Princeton University Press, 1977), 376–77, 33.85–87.

2. Roger Chartier and Peter Stallybrass, "What Is a Book?," in *The Cambridge Companion to Textual Scholarship*, ed. Neil Fraistat and Julia Flanders (Cambridge: Cambridge University Press, 2013), 194.

3. Adrian Johns, *The Nature of the Book: Print and Knowledge in the Making* (Chicago: University of Chicago Press, 2009), 2–3.

4. Robert S. Emmett and David E. Nye, "The Emergence of the Environmental Humanities," in *The Environmental Humanities: A Critical Introduction* (Cambridge, MA: MIT Press, 2017), 6.

BIBLIOGRAPHY

PRIMARY

Alighieri, Dante. *Paradiso*. In *The Divine Comedy*, ed. and trans. Charles S. Singleton, vol. 3, pt. 1. Princeton: Princeton University Press, 1977.

Anonymous. "Adventures of a Quire of Paper." *London Magazine; or Gentlemen's Monthly Intelligencer*, August 1779, 355–58.

Anonymous. "Adventures of a Quire of Paper." *London Magazine; or Gentlemen's Monthly Intelligencer*, October 1779, 448–52.

Anonymous. *The Famovs Victories of Henry the fifth: Containing the Honourable Battell of Agin-court: As it was plaide by the Queenes Maiesties Players*. STC 13072. London: Printed by Thomas Creede, 1598.

Bacon, Francis. *The New Organon*. Ed. Lisa Jardine. Trans. Michael Silverthorne. Cambridge: Cambridge University Press, 2000.

Biblia: The Bible [The Coverdale Bible]. STC 2063. [Cologne: Printed by E. Cervicornus and J. Soter], 1535.

The Byble [The Matthews Bible]. STC 2066. Antwerp: Printed by Matthew Crom for Richard Grafton and Edward Whitchurch, London, 1537.

Castiglione, Baldassare. *The Covrtyer of Covnt Baldessar Castilio, Devided into foure Bookes*. Trans. Thomas Hoby. STC 4779. London: By Henry Denham, 1577. British Library, 8404.ccc.25.

Cavendish, Margaret. "Letter CL." In *CCXI. Sociable Letters*. Wing N872. London: Printed by William Wilson, 1664.

Chaucer, Geoffrey. *The Riverside Chaucer*. Ed. Larry Dean Benson. New York: Oxford University Press, 2008.

Comenius, Johannes Amos. *Orbis Sensualium Pictus*. Trans. Charles Hoole. Wing C5523. London: Printed for J. Kirton, 1659.

Cordier, Mathurin. *Corderius Dialogues Translated Grammatically for the More Speedy Attaining to the Knowledge of the Latine Tongue*. STC 5762. London: Printed by Humfrey Lownes, for Thomas Man, 1614.

Cornwallis, William. *Essayes*. STC 5775. London: Printed [by S. Stafford and R. Read] for Edmund Mattes, 1600–1601.

Cowley, Abraham. *The Guardian, A Comedie Acted before Prince Charls, His Highness at Trinity-Colledg in Cambridge, upon the twelfth of March, 1641*. Wing C6673. London: Printed for John Holden, 1650.

Crofutt, George A. *Crofutt's New Overland Tourist and Pacific Coast Guide, Containing a Condensed and Authentic Description of over One Thousand Two Hundred Cities, Towns, Villages, Stations, Government Fort and Camps, Mountains, Lakes, Rivers, Sulphur, Soda and Hot Springs, Scenery, Watering Places, and Summer Resorts . . . : Over the Union, Central and Southern Pacific Railroads, Their Branches and Connections, by Rail, Water and Stage* [. . .] .Vol. 1. Chicago: Overland, 1878–79. https://hdl.handle.net/2027/nyp.33433081825006?urlappend = %3Bseq = 285.

d'Alembert, Jean-Baptiste le Rond, and Denis Diderot, eds. *Encyclopédie, ou Dictionnaire raisonné des sciences, des arts et des métiers*. Vol. 5. Paris: Briasson et al., 1767.

Dekker, Thomas, and George Wilkins. *Iests to make you Merie: With The Coniuring vp of Cock VVatt, (the walking Spirit of Newgate) To tell Tales*. STC 6541. London: By N[icholas] O[kes] for Nathaniell Butter, dwelling neere to St. Austins Gate, at the signe of the pide Bull, 1607.

Donne, John. *Deaths dvell, or, A Consolation to the Soule, against the dying Life, and liuing Death of the Body. Deliuered in a Sermon at White Hall, before the Kings Maiesty, in the beginning of Lent, 1630. By that late learned and Reuerend Diuine, Iohn Donne, Dr. in Diuinity, & Deane of S. Pauls, London. Being his last Sermon, and called by his Maiesties houshold The Doctors owne Funerall Sermon.* STC 7031. London: Printed by Thomas Harper, for Richard Redmer and Beniamin Fisher, and are to be sold at the signe of the Talbot in Alders-gate street, M.DC.XXXII. [1632].

Donne, John. *The Epithalamions, Anniversaries, and Epicedes.* Ed. W. Milgate. Oxford: Oxford University Press, 1978.

Donne, John. "A Fvneral Elegie." In *An Anatomy of the Vvorld.* STC 7022. London: Printed [by William Stansby] for Samuel Macham, 1611.

Donne, John. *John Donne: The Complete English Poems.* Ed. A. J. Smith. London: Penguin, 1974.

Donne, John. *Letters to Severall Persons of Honour: Written By John Donne.* Wing D1864. London: Printed by J. Flesher for Richard Marriot, 1651.

Donne, John. *Poems, by J. D. with Elegies on the Authors Death.* STC 7045. London: Printed by M. F. for Iohn Marriot, and are to be sold at his shop in St *Dunstans* Church-yard in *Fleet-street*, 1633.

Donne, John. *The Variorum Edition of the Poetry of John Donne.* Vol. 8, *The Epigrams, Epithalamions, Epitaphs, Inscriptions, and Miscellaneous Poems.* Ed. Gary A. Stringer. Bloomington: Indiana University Press, 1995.

Dryden, John. "Palamon and Arcite." In *Fables Ancient and Modern.* Wing D2278. London: Printed for Jacob Tonson, 1700.

Dryden, John. *Religio Laici; or, A Layman's Faith.* Wing D2342. London: Printed for Jacob Tonson, 1682.

Dylan, Bob. "Stuck Inside of Mobile with the Memphis Blues Again." Track 6 on *Blonde on Blonde.* Columbia Records, 1966.

Falkland, Henry Cary. *The Mariage Night Written by the Lord Viscount Fawlkland.* Wing F315. London: printed by W. G[odbid] for R. Crofts at the Crown in Chancery-Lane under Sergeants- Inne, 1664.

Fleischer, Dave. *Any Rags?* New York: Paramount Pictures, 1932.

Florio, John. *A Worlde of Wordes.* STC 11098. London: By Arnold Hatfield for Edw. Blount, 1598.

Freneau, Philip. "On the Emigration to America and Peopling the Western Country" (1785). In *The Norton Anthology of American Literature*, ed. Nina Baym, Ronald Gottesman, Laurence B. Holland, David Kalstone, Francis Murphy, Hershel Parker, William H. Pritchard, and Patricia B. Wallace, 1:719. 3rd ed. New York: W. W. Norton, 1989.

Fuller, Thomas. *Davids Hainovs Sinne.* STC 11463. London: Printed by Tho. Cotes, for Iohn Bellamie, dwelling at the three Golden Lyons in Cornehill, 1631.

Gascoigne, George. *A Hundreth Sundrie Flowres.* Ed. G. W. Pigman III. Oxford: Clarendon Press, 2000.

Gascoigne, George. *A Hundreth sundrie Flowres bounde vp in one small Poesie.* STC 11635. London: Imprinted [by Henrie Bynneman [and Henry Middleton]] for Richarde Smith, 1573. British Library, C.13.a.16.(1).

The Geneva Bible: A Facsimile of the 1560 Edition. Peabody, MA: Hendrickson, 2007.

Gerard, John. *The Herball, or, Generall Historie of Plantes.* STC 11752. London: Printed by Adam Islip, Ioice Norton, and Richard Whitakers, 1636.

Hales, John. *A Compendious or Briefe Examination of Certayne Ordinary Complaints* [. . .]. STC 23133. Imprinted at London: In Fleetstreate, neere vnto Saincte Dunstones Church, by Thomas Marshe, 1581.

Harvey, Richard. *Plaine Percevall the Peace-Maker of England.* STC 12914. London: Printed in Broadstreete at the signe of the Pack-staffe [i.e., Eliot's Court Press for G. Seton], 1590.

Helmont, Jean Baptiste van. *Van Helmont's Works: Containing his most Excellent Philosophy, Physick, Chirurgery, Anatomy.* Trans. J. C. Wing H1397. London: Printed by Lodowick Lloyd, 1664.

The Holie Bible [The Bishops Bible]. STC 2099. London: By Richarde Iugge, 1568.

The Holie Bible [The Douai-Rheims Old Testament]. STC 2207. Doway: By Laurence Kellam, 1609–10.

The Holy Bible [The Geneva Bible]. STC 2129. London: By Christopher Barker, printer to the Queenes most excellent Maiestie, 1580. Folger shelf mark STC 2129, "The Newby Bible" (1580).

The Holy Bible [The King James Bible]. STC 2217. London: By Robert Barker, 1611.

The Holy Bible [The Wicked Bible]. STC 2296. Printed at London: By Robert Barker [. . .], and by the assignes of Iohn Bill, 1631.

Hooke, Robert. *Micrographia, or, Some Physiological Descriptions of Minute Bodies Made by Magnifying Glasses with Observations and Inquiries Thereupon.* Wing H2620. London: Printed by Jo. Martyn and Ja. Allestry, 1665.

Horman, William. *Vulgaria Uiri Doctissimi Guil. Hormani Caesariburgensis.* STC 13811. London: Printed by Richard Pynson, 1519.

Imberdis, Jean. *Papyrus Sive Ars Conficiendae Papyri (Papyrus, or The Craft of Paper).* Trans. Eric Laughton. Hilversum, Holland: The Paper Publications Society, 1952.

Jonson, Ben. "To the memory of my beloued, The Author." In *Mr. VVilliam Shakespeares comedies, histories, & tragedies: published according to the true originall copies.* By William Shakespeare, ed. Iohn Heminge and Henry Condell. STC 22273. London: Printed by Isaac Iaggard, and Ed. Blount, 1623. The Horace Howard Furness Shakespeare Library. Schoenberg Center for Electronic Text & Image (SCETI). http://sceti.library.upenn.edu/sceti/printedbooksNew/index.cfm?TextID=first folio&PagePosition=9.

Koops, Matthias. *Historical Account of the Substances Which Have Been Used to Describe Events, and to Convey Ideas, from the Earliest Date, to the Invention of Paper.* London: T. Burton, 1800.

Koops, Matthias. *Historical Account of the Substances Which Have Been Used to Describe Events, and to Convey Ideas, from the Earliest Date to the Invention of Paper.* 2nd ed. London: Jaques, 1801.

Lalande, Joseph Jérôme Le Français de. *Art de faire le papier. Description des arts et métiers.* Paris: Chez Desaint & Saillant, Libraries [. . .], 1761.

la Lande, Joseph de. *The Art of Making Paper.* Leicestershire: Plough Press, 1978.

Leigh, Edward. *A treatise of divinity consisting of three books.* Wing L1011. London: Printed by E. Griffin for William Lee, 1646.

Maier, Michael. *Lusus Serius: or, Serious Passe-time. A Philosophicall Discourse concerning the Superiority of Creatures under Man.* Trans. J. de la Salle [John Hall]. Wing M286. London: Printed for Humphrey Moseley and Tho. Heath, 1654. Folger Shakespeare Library, 156–125q.

Milton, John. *Of Reformation touching church-discipline in England and the causes that hitherto have hindred it.* Wing M2134. [London]: Printed for Thomas Underhill, 1641.

More, Thomas. *The Confutation of Tyndale's Answer.* Ed. Louis A. Schuster, Richard C. Marius, James P. Lusardi, and Richard J. Schoeck. In *The Complete Works of St. Thomas More,* vol. 8, pt. 1. New Haven: Yale University Press, 1973.

More, Thomas. *A Dialogue Concerning Heresies.* Ed. Thomas M. C. Lawler, Germain P. Marc'hadour, and Richard C. Marius. In *The Complete Works of St. Thomas More,* vol. 6, pt 1. New Haven: Yale University Press, 1981.

More, Thomas. *Memoirs of Sir Thomas More, &c.* Ed. Arthur Cayley, the Younger, Esq. Vol. 2. London: Cadell and Davis, 1808.

More, Thomas. *Utopia.* Ed. George M. Logan, and Robert M. Adams. Trans. Robert M. Adams. Cambridge: Cambridge University Press, 1989.

Moxon, Joseph. *Mechanick Exercises: Or, the Doctrine of Handy-Works.* Wing M3013–3014. Vol. 2. London: Printed for Joseph Moxon, 1677.

Nashe, Thomas. *The Vnfortunate Traueller. Or, The life of Iacke Wilton Newly corrected and augmented.* STC 18381. London: Imprinted by Thomas Scarlet for Cuthbert Burby, 1594.

The New Testament of Iesus Christ [The Douai-Rheims New Testament]. STC 2884. Rhemes: By Iohn Fogny, 1582.

Plat, Sir Hugh. *The Iewell House of Art and Nature Conteining Diuers Rare and Profitable Inuentions, Together with Sundry New Experimentes in the Art of Husbandry.* STC 19991. London: Printed by Peter Short, 1594.

Preston, John. *The Saints Svbmission, and Sathans overthrow.* STC 20266. London: Printed by J. D[awson] and are to be sold by Peter Cole at the signe of the Glove in Cornhill, neere the Royal-Exchange, 1638.

Prynne, William. *Histrio-mastix. The players scourge, or, actores tragædie.* STC 20464a. London: Printed by E[dward] A[llde, Augustine Mathewes, Thomas Cotes] and W[illiam] I[ones] for Michael Sparke, 1633.

Quarles, Francis. *Enchyridion: containing institutions divine contemplative practicall, morall ethycall oeconomicall politicall.* STC 20543. London: Tho. Cotes, 1640.

R., D. *Naaman the Syrian His Disease and Cvre Discovering Lively to the Reader the Spirituall Leprosie of Sinne and Selfe-Love.* Wing R1799. London: Printed by Th. Harper for Philip Nevil, 1642.

Rutherford, Samuel. *Lex, Rex: The Law and the Prince.* Wing R2386. London: Printed for John Field, 1644. Fisher Library, University of Toronto, Forbes 00804.

Schäffer, Jacob Christian. *Versuche und Muster ohne alle Lumpen oder doch mit einem geringen Zusatze derselben Papier zu machen.* Regensburg, 1765.

Shakespeare, William. *Hamlet.* In *Shakespeare's Plays, Sonnets, and Poems* from Folger Digital Texts, ed. Barbara Mowat, Paul Werstine, Michael Poston, and Rebecca Niles. Washington, DC: Folger Shakespeare Library. www.folgerdigitaltexts.org.

Shakespeare, William. *Henry IV, Part 1.* In *Shakespeare's Plays, Sonnets, and Poems* from Folger Digital Texts, ed. Barbara Mowat, Paul Werstine, Michael Poston, and Rebecca Niles. Washington, DC: Folger Shakespeare Library. www.folgerdigitaltexts.org.

Shakespeare, William. *Henry IV, Part 2.* In *Shakespeare's Plays, Sonnets, and Poems* from Folger Digital Texts, ed. Barbara Mowat, Paul Werstine, Michael Poston, and Rebecca Niles. Washington, DC: Folger Shakespeare Library. www.folgerdigitaltexts.org.

Shakespeare, William. *Henry V.* In *Shakespeare's Plays, Sonnets, and Poems* from Folger Digital Texts, ed. Barbara Mowat, Paul Werstine, Michael Poston, and Rebecca Niles. Washington, DC: Folger Shakespeare Library. www.folgerdigitaltexts.org.

Shakespeare, William. *Henry VI, Part 2.* In *Shakespeare's Plays, Sonnets, and Poems* from Folger Digital Texts, ed. Barbara Mowat, Paul Werstine, Michael Poston, and Rebecca Niles. Washington, DC: Folger Shakespeare Library. www.folgerdigitaltexts.org.

Shakespeare, William. *Macbeth.* In *Shakespeare's Plays, Sonnets, and Poems* from Folger Digital Texts, ed. Barbara Mowat, Paul Werstine, Michael Poston, and Rebecca Niles. Washington, DC: Folger Shakespeare Library. www.folgerdigitaltexts.org.

Shakespeare, William. *The Merry Wives of Windsor.* In *Shakespeare's Plays, Sonnets, and Poems* from Folger Digital Texts, ed. Barbara Mowat, Paul Werstine, Michael Poston, and Rebecca Niles. Washington, DC: Folger Shakespeare Library. www.folgerdigitaltexts.org.

Shakespeare, William. *A Midsummer Night's Dream.* In *Shakespeare's Plays, Sonnets, and Poems* from Folger Digital Texts, ed. Barbara Mowat, Paul Werstine, Michael Poston, and Rebecca Niles. Washington, DC: Folger Shakespeare Library. www.folgerdigitaltexts.org.

Shakespeare, William. *Mr. VVilliam Shakespeares Comedies, Histories, & Tragedies: Published according to the True Originall Copies.* Ed. Iohn Heminge and Henry Condell. STC 22273. London: Printed by Isaac Iaggard, and Ed. Blount, 1623.

Shakespeare, William. *Richard II.* In *Shakespeare's Plays, Sonnets, and Poems* from Folger Digital Texts, ed. Barbara Mowat, Paul Werstine, Michael Poston, and Rebecca Niles. Washington, DC: Folger Shakespeare Library. www.folgerdigitaltexts.org.

Shakespeare, William. *Richard III.* In *Shakespeare's Plays, Sonnets, and Poems* from Folger Digital Texts, ed. Barbara Mowat, Paul Werstine, Michael Poston, and Rebecca Niles. Washington, DC: Folger Shakespeare Library. www.folgerdigitaltexts.org.

Shakespeare, William. *Sonnets.* In *Shakespeare's Plays, Sonnets, and Poems* from Folger Digital Texts, ed. Barbara Mowat, Paul Werstine, Michael Poston, and Rebecca Niles. Washington, DC: Folger Shakespeare Library. www.folgerdigitaltexts.org.

Shute, Nathaniel. *Corona Charitatis, The Crowne of Charitie: A Sermon Preacht in Mercers Chappell, May 10. 1625.* STC 22466. London: Printed by W. Stansby for Samuel Man, dwelling at the Swanne in Pauls Church-yard, 1626.

Sidney, Sir Philip. *Sir Philip Sidney's an Apology for Poetry, And, Astrophil and Stella: Texts and Contexts.* Ed. Peter C. Herman. Glen Allen, VA: College Publishing, 2001.

Sparke, Michael. *Scintilla, or, a light broken into darke Warehouses.* Wing S4818B. London: Printed, not for profit, but for the common weles good: and no where to be sold, but some where to be given, 1641.

Spenser, Edmund. *The Faerie Queene.* Ed. A. C. Hamilton with Hiroshi Yamashita, Toshiyuki Suzuki, and Shohachi Fukuda. 2nd ed. London: Routledge, 2001.

Stow, John. *The Survey of London: Contayning the Originall, Increase, Moderne Estate, and Government of that City; Methodically Set downe.* STC 23345. London: Printed by Elizabeth Pvrslovv, and to be sold by Nicholas Bovrne, 1633.

Tatham, John. *The Scots Figgaries: or, A Knot of Knaves.* Wing T235. London: Printed by W. H. for Iohn Tey, 1652.
Taylor, John. *The Praise of Hemp-seed.* STC 23788. London: For H. Gosson, 1620.
Thoreau, Henry David. *Walden.* Ed. Jeffrey S. Cramer. New Haven: Yale University Press, 2006.
Vaughan, Henry. *Silex Scintillans: Sacred Poems and Private Ejaculations.* 2nd ed. Wing V126. London: Printed for Henry Crips and Lodowick Lloyd, 1655.
Vicars, John. *Unholsome henbane between two fragrant roses.* Wing V332. London: Printed for John Rothwell, 1645.
Vives, Juan Luis. *The Instruction of a Christen Woman.* Ed. Virginia Walcott Beauchamp, Elizabeth H. Hageman, and Margaret Mikesell. Urbana: University of Illinois Press, 2002.
Wilder, Laura Ingalls. *Farmer Boy.* New York: HarperTrophy, 2007.

SECONDARY

"£30 Million Investment in a Sustainable Future." National Trust. August 8, 2016. https://www.nationaltrust.org.uk/news/30-million-investment-in-a-sustainable-future.
"About." *Lost Books: Reconstructing the Print World of Pre-Industrial Europe.* Brill. March 21, 2016. https://brill.com/abstract/title/32814.
"About the First Folio Tour." Folger Shakespeare Library. Accessed December 10, 2018. http://www.folger.edu/about-the-first-folio-tour.
"About the Study—Microbiomes of the Built Environment." The National Academies of Sciences, Engineering, and Medicine. Accessed August 18, 2017. http://nas-sites.org/builtmicrobiome/about-the-study-2.
"About Us: Crane Currency." Crane and Co. Accessed August 15, 2017. http://www.crane.com/about-us/crane-currency. Available through the Internet Archive Wayback Machine. https://web.archive.org/web/20171007030013/http://www.crane.com:80/about-us/crane-currency.
Acheson, Katherine. "The Occupation of the Margins: Writing, Space, and Early Modern Women." In *Early Modern English Marginalia*, ed. Katherine Acheson, 70–89. Material Readings in Early Modern Culture. London: Routledge, 2018.
Adam, Paul. *Das Restuarieren alter Bücher.* Halle (Saale): Verlag Wilhelm Knapp, 1927.
Albro, Sylvia Rodgers. *Fabriano: City of Medieval and Renaissance Papermaking.* New Castle, DE: Oak Knoll Press, 2016.
Allen, Graham, Carrie Griffin, and Mary O'Connell. Introduction to *Readings on Audience and Textual Materiality*, 1–8. Ed. Graham Allen, Carrie Griffin, and Mary O'Connell. London: Routledge, 2015.
Allsopp, Dennis, Kenneth Seal, and Christine Gaylarde. *Introduction to Biodeterioration.* 2nd ed. Cambridge: Cambridge University Press, 2004.
Antonetti, Martin. "Exploring the Archaeology of the Book in the Liberal Arts Curriculum." In *Teaching Bibliography, Textual Criticism, and Book History*, ed. Ann R. Hawkins, 19–24. London: Pickering & Chatto, 2006.
Arsenault, Raymond. "The End of the Long Hot Summer: The Air Conditioner and Southern Culture." *Journal of Southern History* 50, no. 4 (November 1984): 597–628.
Asunción, Josep. *The Complete Book of Papermaking.* New York: Lark Books, 2003.
Badger, Emily, and Alan Blinder. "How Air-Conditioning Conquered America (Even the Pacific Northwest)." *New York Times*, August 4, 2017.
Bailey, Janette-Susan. *Dust Bowl: Depression America to World War Two Australia.* New York: Palgrave Macmillan, 2016.
Baker, Cathleen A. *From the Hand to the Machine: Nineteenth-Century American Paper and Mediums; Technologies, Materials, and Conservation.* Ann Arbor, MI: The Legacy Press, 2010.
Baker, Cathleen. "Nineteenth-Century Papers: Sizing Does Matter." *Hand Papermaking* 30, no. 2 (Winter 2015): 19–21.
Barcham Green, Simon. "Sizing at Hayle Mill." *Hand Papermaking* 30, no. 2 (Winter 2015): 11–17.
Barrett, Timothy. "Aesthetics and the Future of the Craft." *Hand Papermaking* 11, no. 2 (Winter 1996): 13–17.

Barrett, Timothy. "European Papermaking Techniques, 1300–1800." Paper Through Time: Nondestructive Analysis of 14th- Through 19th-Century Papers. The University of Iowa. Last modified July 31, 2018. http://paper.lib.uiowa.edu/european.php.

Barrett, Timothy. "Listening to Paper." *Hand Papermaking* 32, no. 2 (Winter 2017): 6.

Barrett, Timothy. "Sizing Terminology and Practice." *Hand Papermaking* 30, no. 2 (Winter 2015): 3–6.

Barrett, Timothy, and Cynthea Mosier. "The Role of Gelatin in Paper Permanence." *Journal of the American Institute for Conservation* 34, no. 3 (Autumn–Winter 1995): 173–86.

Barrett, Timothy, et al. "Conclusions." Paper Through Time: Nondestructive Analysis of 14th- Through 19th-Century Papers. The University of Iowa. Last modified July 1, 2013. http://paper.lib.uiowa.edu/conclusions.php.

Barrett, Timothy, et al. "Discussions: Chronological Plots." Paper Through Time: Nondestructive Analysis of 14th- Through 19th-Century Papers. The University of Iowa. Last modified Jan. 5, 2016. http://paper.lib.uiowa.edu/chron.php.

Barrett, Timothy, et al. "Discussions: Non-Chronological Plots." Paper Through Time: Nondestructive Analysis of 14th- Through 19th-Century Papers. The University of Iowa. Last modified January 18, 2012. http://paper.lib.uiowa.edu/nonchron.php.

Barrett, Timothy, et al. Paper Through Time: Nondestructive Analysis of 14th- Through 19th-Century Papers. The University of Iowa. Last modified May 4, 2016. http://paper.lib.uiowa.edu/index.php.

Barrett, Timothy, et al. "Project Goals." Paper Through Time: Nondestructive Analysis of 14th- Through 19th-Century Papers. The University of Iowa. Last modified November 28, 2011. http://paper.lib.uiowa.edu/goals.php.

Barrett, Timothy, et al. "Results: Plot Library." Paper Through Time: Nondestructive Analysis of 14th- Through 19th-Century Papers. The University of Iowa. Last modified January 25, 2012. http://paper.lib.uiowa.edu/plot.php.

Basbanes, Nicholas A. *On Paper: The Everything of Its Two-Thousand-Year History.* New York: Knopf, 2013. Reprint, New York: Vintage, 2014.

Bate, Jonathan. *The Song of the Earth.* Cambridge, MA: Harvard University Press, 2000.

Baty, John. "Sizing Science: Can Physical Science and Engineering Answer Some of Our Lingering Questions?" *Hand Papermaking* 30, no. 2 (Winter 2015): 7–10.

Beal, Peter. *A Dictionary of English Manuscript Terminology, 1450–2000.* Oxford: Oxford University Press, 2008.

Bergthaller, Hannes, Rob Emmett, Adeline Johns-Putra, Agnes Kneitz, Susanna Lidström, Shane McCorristine, Isabel Pérez Ramos, Dana Phillips, Kate Rigby, and Libby Robin. "Mapping Common Ground: Ecocriticism, Environmental History, and the Environmental Humanities." *Environmental Humanities* 5, no. 1 (2014): 261–76.

Berry, Wendell. "'Leadership from the Bottom': Wendell Berry on Rural Revival." *New Atlantis*, no. 21 (2008): 127–29.

Bidwell, John. "French Paper in English Books." In *The Cambridge History of the Book in Britain*, vol. 4, *1557–1695*, ed. John Barnard and D. F. McKenzie, 583–601. Cambridge: Cambridge University Press, 2002.

Blackwell, Mark. *The Secret Life of Things: Animals, Objects, and It-Narratives in Eighteenth-Century England.* Lewisburg: Bucknell University Press, 2007.

Blair, Ann. "The Rise of Note-Taking in Early Modern Europe." *Intellectual History Review* 20, no. 3 (2010): 303–16.

Blair, Ann M. *Too Much to Know: Managing Scholarly Information Before the Modern Age.* New Haven: Yale University Press, 2010.

Bland, Mark. *A Guide to Early Printed Books and Manuscripts.* Hoboken, NJ: Wiley, 2010.

Blinde, Loren M. "Rumored History in Shakespeare's *2 Henry IV.*" *English Literary Renaissance* 38, no. 1 (Winter 2008): 34–54.

Bloom, Jonathan M. *Paper Before Print: The History and Impact of Paper in the Islamic World.* New Haven: Yale University Press, 2001.

Bloy, Colin H. *A History of Printing Ink, Balls, and Rollers, 1440–1850.* 1967. Reprint, London: The Wynkyn de Worde Society, 1972.

Borland, Jennifer. "Unruly Reading: The Consuming Role of Touch in the Experience of a Medieval Manuscript." In *Scraped, Stroked, and Bound: Materially Engaged Readings of Medieval Manuscripts*, ed. Jonathan Wilcox, 97–114. Turnhout: Brepols, 2013.

Brayman, Heidi. *Reading Material in Early Modern England: Print, Gender, and Literacy*. Cambridge: Cambridge University Press, 2005.

Brown, Michelle, and Patricia Lovett. *The Historical Source Book for Scribes*. Toronto: University of Toronto Press, 1999.

Brown, Piers. "Donne's Hawkings." *Studies in English Literature, 1500–1900* 49, no. 1 (Winter 2009): 67–86.

Browning, B. L. "The Nature of Paper." *Library Quarterly: Information, Community, Policy* 40, no. 1 (January 1970): 18–38.

Bruni, Flavia, and Andrew Pettegree, eds. *Lost Books: Reconstructing the Print World of Pre-Industrial Europe*. Leiden: Brill, 2016.

Bulman, James C., ed. *King Henry IV Part 2*. By William Shakespeare. London: Bloomsbury Arden Shakespeare, 2016.

Buss, James Joseph. *Winning the West with Words: Language and Conquest in the Lower Great Lakes*. Norman: University of Oklahoma Press, 2013.

Calhoun, Joshua. "Ecosystemic Shakespeare: Vegetable Memorabilia in the *Sonnets*." *Shakespeare Studies* 39 (2011): 64–73.

Calhoun, Joshua. "The Word Made Flax: Cheap Bibles, Textual Corruption, and the Poetics of Paper." *PMLA* 126, no. 2 (March 2011): 327–44.

Caneva, Giulia, Maria Pia Nugari, and Ornella Salvadori, eds. *Plant Biology for Cultural Heritage: Biodeterioration and Conservation*. Trans. Helen Glanville. Los Angeles: Getty Conservation Institute, 2008.

Carrier, Willis H. "The Economics of Man-Made Weather." *Scientific American* 148, no. 4 (April 1933): 199–202.

Carter, John, and Nicolas Barker. *ABC for Book Collectors*. 8th ed. New Castle, DE: Oak Knoll Press, 2004.

Chartier, Roger, and Peter Stallybrass. "What Is a Book?" In *The Cambridge Companion to Textual Scholarship*, ed. Neil Fraistat and Julia Flanders, 188–204. Cambridge: Cambridge University Press, 2013.

Christopher, Georgia B. "In Arcadia, Calvin . . . : A Study of Nature in Henry Vaughan." In *Essential Articles for the Study of Henry Vaughan*, ed. Alan Rudrum, 177–88. Hamden, CT: Archon Books, 1987.

Clanchy, M. T. "Parchment and Paper Manuscript Culture, 1100–1500." In *A Companion to the History of the Book*, ed. Jonathan Rose and Simon Eliot, 194–206. Chichester, West Sussex: Wiley-Blackwell, 2009.

Clapp, B. W. *An Environmental History of Britain Since the Industrial Revolution*. London: Routledge, 2014.

Clapp, Verner W. "The Story of Permanent/Durable Book Paper, 1115–1970." *Restaurator* 1, no. 3 (1972): 1–51.

Clarkson, Christopher. "Rediscovering Parchment: The Nature of the Beast." *Paper Conservator* 16 (1992): 5–26.

Cockayne, Emily. *Hubbub: Filth, Noise, and Stench in England, 1600–1770*. New Haven: Yale University Press, 2007.

Coleman, D. C. *The British Paper Industry, 1495–1860: A Study in Industrial Growth*. Oxford: Clarendon Press, 1958.

Connell, Alyssa, and Marissa Nicosia. Cooking in the Archives: Updating Early Modern Recipes (1600–1800) in a Modern Kitchen. Last modified September 30, 2018. https://rarecooking.com.

Conrad, Ernest A. "Climate Control Systems Design and Climate Change." Discussion paper for the Getty Conservation Institute: Alternative Climate Controls for Historic Buildings, Tenerife, Spain, April 2007. http://www.getty.edu/conservation/our_projects/science/climate/paper_conrad.pdf.

Cooper, Gail. *Air-Conditioning America: Engineers and the Controlled Environment, 1900–1960*. Baltimore: Johns Hopkins University Press, 2002.

Cope, Wendy. *Making Cocoa for Kingsley Amis*. London: Faber and Faber, 1986.

Cormack, Bradin, and Carla Mazzio. *Book Use, Book Theory, 1500–1700*. Chicago: University of Chicago Library, 2005.

Costley, Clare L. "David, Bathsheba, and the Penitential Psalms." *Renaissance Quarterly* 57, no. 4 (Winter 2004): 1235–77.

Crowley, Lara M. "Archival Research." In *The Oxford Handbook of John Donne*, ed. Dennis Flynn, M. Thomas Hester, and Jeanne Shami, 34–42. Oxford Handbooks Online. September 2012. www.oxfordhandbooks.com.

Dabrowski, Józef, and John S. G. Simmons. "Permanence of Early European Hand-Made Papers." *Fibers and Textiles in Eastern Europe* 11, no. 1 (March 2003): 8–13.

Dane, Joseph A. *Abstractions of Evidence in the Study of Manuscripts and Early Printed Books*. Farnham: Ashgate, 2009.

Dane, Joseph. *What Is a Book? The Study of Early Printed Books*. Notre Dame: University of Notre Dame Press, 2012.

Daniell, David. *William Tyndale: A Biography*. 1994. Reprint, New Haven: Yale University Press, 2001.

Darnton, Robert. "What Is the History of Books?" *Daedalus* 111, no. 3 (Summer 1982): 65–83.

Darnton, Robert. " 'What Is the History of Books?' Revisited." *Modern Intellectual History* 4, no. 3 (2007): 495–508.

Davies, Keri. "William Blake and the Straw Paper Manufactory at Millbank." In *Blake in Our Time: Essays in Honour of G. E. Bentley Jr.*, ed. Karen Mulhallen, 233–61. Toronto: University of Toronto Press, 2010.

Davis, Charles Thomas. *The Manufacture of Paper: Being a Description of the Various Processes for the Fabrication, Coloring, and Finishing of Every Kind of Paper*. Philadelphia: Henry Carey Baird, 1886.

de Grazia, Margreta. "Words as Things." *Shakespeare Studies* 28 (2000): 231–35.

de Grazia, Margreta, and Peter Stallybrass. "The Materiality of the Shakespearean Text." *Shakespeare Quarterly* 44, no. 3 (Fall 1993): 255–83.

De Vinne, Theodore L. *The Invention of Printing: A Collection of Facts and Opinions*. New York: F. Hart, 1878.

Donaldson, Ian. "Jonson, Benjamin [Ben] (1572–1637), poet and playwright." *Oxford Dictionary of National Biography*. Accessed July 21, 2017. http://www.oxforddnb.com/view/article/15116.

Douglas, Ian. *Cities: An Environmental History*. New York: I.B. Tauris, 2013.

Drake, James David. *The Nation's Nature: How Continental Presumptions Gave Rise to the United States of America*. Charlottesville: University of Virginia Press, 2011.

Duarte, Marisa Elena. *Network Sovereignty: Building the Internet across Indian Country*. Seattle: University of Washington Press, 2017.

Echard, Siân. "House Arrest: Modern Archives, Medieval Manuscripts." *Journal of Medieval and Early Modern Studies* 30, no. 2 (Spring 2000): 185–210.

Ellis, Erle C., and Navin Ramankutty. "Putting People in the Map: Anthropogenic Biomes of the World." *Frontiers in Ecology and the Environment* 6, no. 8 (October 2008): 439–47.

Ellis, Margaret Holben, ed. *Historical Perspectives in the Conservation of Works of Art on Paper*. Los Angeles: Getty Publications, 2015.

Emmett, Robert S., and David E. Nye. "The Emergence of the Environmental Humanities." In *The Environmental Humanities: A Critical Introduction*, 1–21. Cambridge, MA: MIT Press, 2017.

Enniss, Stephen C., and Erin C. Blake. "Folger Sustainable Preservation Environment Project White Paper." Folger Shakespeare Library. November 30, 2011. https://hcommons.org/deposits/item/hc:12487.

Estok, Simon C. "Afterword: Ecocriticism on the Lip of a Lion." In *Ecocritical Shakespeare*, ed. Lynne Bruckner and Dan Brayton, 239–46. Farnham: Ashgate, 2011.

Estok, Simon C. *Ecocriticism and Shakespeare: Reading Ecophobia*. New York: Palgrave Macmillan, 2011.

European Commission. "Data Protection in the EU, Article 17: 'Right to be forgotten and to erasure.' " *Proposal for a Regulation of the European Parliament and of the Council on the protection of individuals with regard to the processing of personal data and on the free movement of such data (General Data Protection Regulation)*. Accessed June 14, 2018. https://ec.europa.eu/info/law/law-topic/data-protection/data-protection-eu_en.

Evans, Meredith. "Rumor, the Breath of Kings, and the Body of Law in *2 Henry IV*." *Shakespeare Quarterly* 60, no. 1 (Spring 2009): 1–24.

"Experts' Roundtable on Sustainable Climate Management Strategies." Roundtable for the Getty Conservation Institute: Alternative Climate Controls for Historic Buildings, Tenerife, Spain, April 2007. http://www.getty.edu/conservation/our_projects/science/climate/roundtable_transcript.pdf.

"Explore the Cathedral: South Quire Aisle." St. Paul's Cathedral. Accessed September 25, 2016. https://www.stpauls.co.uk/history-collections/history/explore-the-cathedral.

Farmer, Alan B. "Playbooks and the Question of Ephemerality." In *The Book in History, the Book as History: New Intersections of the Material Text*, ed. Heidi Brayman, Jesse M. Lander, and Zachary Lesser, 87–125. New Haven: Beinecke Rare Book & Manuscript Library, Yale University, 2016.

Febvre, Lucien, and Henri-Jean Martin. *The Coming of the Book: The Impact of Printing, 1450–1800*. Ed. Geoffrey Nowell-Smith and David Wooton. Trans. David Gerard. London: Verso, 1997.

Feerick, Jean E., and Vin Nardizzi, eds. *The Indistinct Human in Renaissance Literature*. New York: Palgrave Macmillan, 2012.

Felski, Rita. *Uses of Literature*. Hoboken, NJ: Wiley-Blackwell, 2008.

Ferrell, Lori Anne. *The Bible and the People*. New Haven: Yale University Press, 2008.

Finlay, Michael. *Western Writing Implements in the Age of the Quill Pen*. Carlisle, UK: Plains Books, 1990.

Flibbert, Nicole. "The 'Art of Scraping': Knife Erasures in Seventeenth-Century English Manuscript Plays." *Journal of the Penn Manuscript Collective* 1, no. 1 (August 1, 2017). https://repository.upenn.edu/manuscript_collective_journal/vol1/iss1/3.

Folger Shakespeare Library. "The Praise of Hemp-seed: John Taylor References Shakespeare by Name." Shakespeare Documented. May 25, 2017. http://www.shakespearedocumented.org/exhibition/document/praise-hemp-seed-john-taylor-references-shakespeare-name.

From Papyri to King James: The Transmission of the English Bible. University of Michigan Library. 2004. https://www.lib.umich.edu/files/collections/papyrus/exhibits/papyri_james/review.html.

Frontain, Raymond-Jean. "Donne's Imperfect Resurrection." *Papers on Language & Literature* 26, no. 4 (Fall 1990): 539–45.

Galey, Alan. *The Shakespearean Archive: Experiments in New Media from the Renaissance to Postmodernity*. Cambridge: Cambridge University Press, 2014.

Garlick, Karen. "A Brief Review of the History of Sizing and Resizing Practices." *Book and Paper Group Annual* 5 (1986). cool.conservation-us.org.

Gaskell, Philip. *A New Introduction to Bibliography*. New Castle, DE: Oak Knoll Press, 2009.

Gitelman, Lisa. *Always Already New: Media, History, and the Data of Culture*. Cambridge, MA: MIT Press, 2008.

Gitelman, Lisa. *Paper Knowledge: Toward a Media History of Documents*. Durham: Duke University Press, 2014.

Goulden, R. J. "Koops, Matthias (*fl.* 1789–1805)." *Oxford Dictionary of National Biography*. Accessed August 3, 2015. http://www.oxforddnb.com/view/article/37644.

Green, Ian. *Print and Protestantism in Early Modern England*. Oxford: Oxford University Press, 2000.

Green, Jonathan, Frank McIntyre, and Paul Needham. "The Shape of Incunable Survival and Statistical Estimation of Lost Editions." *Papers of the Bibliographic Society of America (PBSA)* 105, no. 2 (June 2011): 141–75.

Greenberg, Amy S. *Manifest Manhood and the Antebellum American Empire*. Cambridge: Cambridge University Press, 2005.

Greetham, D. C. *Textual Scholarship: An Introduction*. New York: Garland, 1994.

Greg, W. W. "The Rationale of Copy-Text." *Studies in Bibliography* 3 (1950): 19–36.

Haines, John. "Erasures in Thirteenth-Century Music." In *Music and Medieval Manuscripts: Paleography and Performance; Essays Dedicated to Andrew Hughes*, ed. Andrew Hughes, John Dickinson Haines, and Randall Rosenfeld, 60–88. Burlington, VT: Ashgate, 2004.

Hamady, Walter. "Notes Regarding the Activity of Papermaking by Hand." In *Papermaking*, by Jules Heller. New York: Watson-Guptill, 1978.

Hamady, Walter. *Papermaking by Hand: A Book of Suspicions*. Mount Horeb, WI: Perishable Press, 1982.

Hamlin, Hannibal. *The Bible in Shakespeare*. Oxford: Oxford University Press, 2013.

Hamlin, Hannibal. *Psalm Culture and Early Modern English Literature*. Cambridge: Cambridge University Press, 2004.

Harris, Jonathan Gil. *Untimely Matter in the Time of Shakespeare*. Philadelphia: University of Pennsylvania Press, 2009.

Harvey, Elizabeth D. "Ventriloquizing Sappho: Ovid, Donne, and the Erotics of the Feminine Voice." *Criticism* 31, no. 2 (Spring 1989): 115–38.

Hawkins, Peter S. *Dante's Testaments: Essays in Scriptural Imagination.* Stanford: Stanford University Press, 1999.

Hayles, N. Katherine. "The Complexities of Seriation." In "Mobile Citizens, Media States," special issue, *PMLA* 117, no. 1 (January 2002): 117–21.

Hayles, N. Katherine. *How We Became Posthuman: Virtual Bodies in Cybernetics, Literature, and Informatics.* Chicago: University of Chicago Press, 1999.

Hayles, N. Katherine. *Writing Machines.* Cambridge: MIT Press, 2002.

Hayles, N. Katherine, and MIT Press. "Writing Machines Web Supplement." MIT.edu. https://mitpress.mit.edu/sites/default/files/titles/content/mediawork/titles/writing/sup/sup_index.html.

Hector, L. C. *The Handwriting of English Documents.* 2nd ed. London: Edward Arnold, 1966.

Hedges, S. Blair. "Wormholes Record Species History in Space and Time." *Biology Letters* 9 (2012): 1–4.

Heffernan, Megan. "Turning Sonnets into Poems: Textual Affect and John Benson's Metaphysical Shakespeare." *Shakespeare Quarterly* 64, no. 1 (Spring 2013): 71–98.

Heise, Ursula K. "Introduction: Planet, Species, Justice—and the Stories We Tell About Them." In *The Routledge Companion to the Environmental Humanities*, ed. Ursula K. Heise, Jon Christensen, and Michelle Niemann, 1–10. London: Routledge, 2017.

Hellinga, Lotte. "Printing." In *The Cambridge History of the Book in Britain*, vol. 3, *1400–1557*, ed. Lotte Hellinga and J. B. Trapp, 91–97. Cambridge: Cambridge University Press, 1999.

Henderson, Ernest F. *Select Historical Documents of the Middle Ages.* London: George Bell and Sons, 1896.

Hester, M. Thomas. "'Miserrimum dictu': Donne's Epitaph for His Wife." *Journal of English and Germanic Philology* 94, no. 4 (October 1995): 513–29.

Hester, M. Thomas, Robert Parker Sorlien, and Dennis Flynn, eds. *John Donne's Marriage Letters in the Folger Shakespeare Library.* Washington, DC: Folger Shakespeare Library, Distributed by University of Washington Press, 2005.

Higham, Robert R. A. *A Handbook of Papermaking.* London: Oxford University Press, 1963.

Hill, Alexandra. "Lost Print in England: Entries in the Stationers' Company Register, 1557–1640." In *Lost Books: Reconstructing the Print World of Pre-Industrial Europe*, ed. Flavia Bruni and Andrew Pettegree, 144–59. Leiden: Brill, 2016.

Hills, Richard L. *Papermaking in Britain, 1488–1988: A Short History.* London: The Athlone Press, 1988.

Hills, Richard. "Some Notes on the Matthias Koops Papers." *The Quarterly: The Review of British Association of Paper Historians*, no. 2 (August 1990): 3–5.

Hodgdon, Barbara. "Historiography and the Uses of History." In *The First Part of King Henry the Fourth: Texts and Contexts*, ed. Barbara Hodgdon, 121–68. Boston: Bedford Books, 1997.

Holsinger, Bruce. "Of Pigs and Parchment: Medieval Studies and the Coming of the Animal." *PMLA* 124, no. 2 (March 2009): 616–23.

Hooks, Adam G. "Making Histories; or, Shakespeare's *Ring*." In *The Book in History, the Book as History: New Intersections of the Material Text*, ed. Heidi Brayman, Jesse M. Lander, and Zachary Lesser, 341–73. New Haven: Beinecke Rare Book & Manuscript Library, Yale University, 2016.

Hooks, Adam G. *Selling Shakespeare: Biography, Bibliography, and the Book Trade.* Cambridge: Cambridge University Press, 2016.

Howsam, Leslie. *Old Books and New Histories: An Orientation to Studies in Book and Print Culture.* Toronto: University of Toronto Press, 2006.

Howsam, Leslie. "The Study of Book History." In *The Cambridge Companion to the History of the Book*, ed. Leslie Howsam, 1–13. Cambridge: Cambridge University Press, 2014.

Hueck, H. J. "The Biodeterioration of Materials as Part of Hylobiology." *Material und Organismen* 1 (1965): 5–34.

Hunter, Dard. *Papermaking: The History and Technique of an Ancient Craft.* New York: Dover, 1947.

Hunter, Dard. *Papermaking in Pioneer America.* Philadelphia: University of Pennsylvania Press, 1952.

Hutchby, Ian. "Technologies, Texts, and Affordances." *Sociology* 35, no. 2 (2001): 441–56.

Hutchinson, Francis Ernest. *Henry Vaughan: A Life and Interpretation.* Oxford: Clarendon Press, 1947.

Ioppolo, Grace. *Revising Shakespeare.* Cambridge, MA: Harvard University Press, 1991.

Iyengar, Sujata. "Intermediating the Book Beautiful: Shakespeare at the Doves Press." *Shakespeare Quarterly* 67, no. 4 (2016): 481–502.

Jacobson, Miriam. "Vibrant Hair: Temporality and the Undead in Early Modern Poetry." Paper presented at Folger Shakespeare Library, Washington, DC, Summer 2014.

Johns, Adrian. *The Nature of the Book: Print and Knowledge in the Making*. Chicago: University of Chicago Press, 2009.

Johnston, Ruth A. *All Things Medieval: An Encyclopedia of the Medieval World*. Vol. 1, A–I. Santa Barbara: Greenwood, 2011.

Jones, Ann Rosalind, and Peter Stallybrass. *Renaissance Clothing and the Materials of Memory*. Cambridge: Cambridge University Press, 2000.

Jones, Meg Leta. *Ctrl + Z: The Right to Be Forgotten*. New York: NYU Press, 2018.

"Joseph Jerome de la Lande, *Art de Faire le Papier* (Paris, France, 1761)." Cary Graphic Arts Collection Image Database. Accessed January 16, 2017. http://library.rit.edu/cary/joseph-jerome-de-la-lande-art-de-faire-le-papier.

"Journal: Autograph Manuscript Signed: Walden [Concord], 1845 July 5–1846 Mar. 27." The Morgan Library & Museum. Accessed July 25, 2017. https://www.themorgan.org/literary-historical/288111.

Jurkiewicz, Iwona. "Magic in Conservation—Using Leaf-Casting on Paper and Palm Leaves." *Collection Care* (British Library blog), October 10, 2017. http://blogs.bl.uk/collectioncare/2017/10/magic-in-conservation-using-leaf-casting-a-mechanical-pulp-repair-technique-on-paper-and-palm-leaves-as-the-library-i.html.

Kay, Sarah. "Original Skin: Flaying, Reading, and Thinking in the Legend of Saint Bartholomew and Other Works." *Journal of Medieval and Early Modern Studies* 36, no. 1 (Winter 2006): 35–74.

Ketelaar, Eric. "Archival Temples, Archival Prisons: Modes of Power and Protection." *Archival Science* 2, no. 3 (2002): 221–38.

Knight, Jeffrey Todd. *Bound to Read: Compilations, Collections, and the Making of Renaissance Literature*. Philadelphia: University of Pennsylvania Press, 2013.

Knight, Jeffrey Todd. "'Furnished' for Action: Renaissance Books as Furniture." *Book History* 12 (2009): 37–73.

Knight, Jeffrey Todd. "Making Shakespeare's Books: Assembly and Intertextuality in the Archives." *Shakespeare Quarterly* 60, no. 3 (Fall 2009): 304–40.

Kolbert, Elizabeth. *The Sixth Extinction: An Unnatural History*. New York: Henry Holt, 2014.

Krill, John. *English Artists' Paper: Renaissance to Regency*. 2nd ed. New Castle, DE: Oak Knoll Press and Winterthur Museum, 2002.

Kurlansky, Mark. *Paper: Paging Through History*. New York: W. W. Norton, 2016. Reprint, New York: W. W. Norton, 2017.

Lamb, Jonathan. *The Things Things Say*. Princeton: Princeton University Press, 2011.

Lears, T. J. Jackson. *Fables of Abundance: A Cultural History of Advertising in America*. New York: Basic Books, 1995.

Lemire, Elise. *Black Walden: Slavery and Its Aftermath in Concord, Massachusetts*. Philadelphia: University of Pennsylvania Press, 2009.

Lesser, Zachary. *"Hamlet" After Q1: An Uncanny History of the Shakespearean Text*. Philadelphia: University of Pennsylvania Press, 2015.

Lewalski, Barbara Kiefer. *Protestant Poetics and the Seventeenth-Century Religious Lyric*. Princeton: Princeton University Press, 1979.

Litvack, Emily. "First Folio Fights Off Agents of Decay." Research, Discovery, & Innovation. The University of Arizona. March 7, 2016. https://research.arizona.edu/stories/first-folio-fights-agents-decay.

Lupton, Christina. "Gender and Materiality on the Eighteenth-Century Page." *SEL Studies in English Literature 1500–1900* 54, no. 3 (Summer 2014): 605–24.

Lupton, Christina. "The Knowing Book: Authors, It-Narratives, and Objectification in the Eighteenth Century." *NOVEL* 39, no. 3 (Summer 2006): 402–20.

Lyall, R. J. "Materials: The Paper Revolution." In *Book Production and Publishing in Britain, 1375–1475*, ed. Jeremy Griffiths and Derek Pearsall, 11–29. Cambridge: Cambridge University Press, 1989.

Mahnert, Alexander, Christine Moissl-Eichinger, and Gabriele Berg. "Microbiome Interplay: Plants Alter Microbial Abundance and Diversity Within the Built Environment." *Frontiers in Microbiology* 6 (2015): 1–11.

Mahood, M. M. "Vaughan: The Symphony of Nature." In *Essential Articles for the Study of Henry Vaughan*, ed. Alan Rudrum, 5–45. Hamden: Archon Books, 1987.

"Major Biomes of the World." World Wildlife Fund. Accessed December 10, 2018. http://wwf.panda.org/about_our_earth/teacher_resources/webfieldtrips/major_biomes.

Marc'hadour, Germain P., and Thomas M. C. Lawler. "Scripture in the *Dialogue.*" In *A Dialogue Concerning Heresies*, in *The Complete Works of St. Thomas More*, vol. 6, pt. 2, ed. Thomas M. C. Lawler, Germain P. Marc'hadour, and Richard C. Marius, 494–525. New Haven: Yale University Press, 1981.

Maret, Russell. "Hungry Bibliophiles: An Experiment in Utilitarian Bookmaking." *Hand Papermaking* 30, no. 2 (Winter 2015): 29–35.

Marzi, Flavio. "Fragments in Bindings." *Collection Care* (British Library blog). May 5, 2016. http://blogs.bl.uk/collectioncare/2016/05/fragments-in-bindings.html.

McColley, Diane Kelsey. *Poetry and Ecology in the Age of Milton and Marvell.* Burlington, VT: Ashgate, 2007.

McCullough, Peter. "Donne to Death." The Bodleian Libraries (BODcasts). May 13, 2016. https://podcasts.ox.ac.uk/donne-death.

McKerrow, Ronald B. *An Introduction to Bibliography for Literary Students.* Oxford: Clarendon Press, 1928.

McKenzie, D. F. *Bibliography and the Sociology of Texts.* 1986. Reprint, Cambridge: Cambridge University Press, 1999.

McKenzie, D. F. "Typography and Meaning: The Case of William Congreve." In *Making Meaning: "Printers of the Mind" and Other Essays*, ed. Peter D. McDonald and Michael F. Suarez, 198–236. Amherst: University of Massachusetts Press, 2002.

McKerrow, Ronald B. *An Introduction to Bibliography for Literary Students.* Oxford: Clarendon Press, 1928.

McKitterick, David. *A History of Cambridge University Press.* Vol. 1, *Printing and the Book Trade in Cambridge, 1534–1698.* Cambridge: Cambridge University Press, 1992.

McKitterick, David. *Print, Manuscript, and the Search for Order, 1450–1830.* Cambridge: Cambridge University Press, 2003.

McLeod, Randall. "Obliterature: Reading a Censored Text of Donne's 'To his mistress going to bed.'" *English Manuscript Studies* 12 (2005): 83–138.

McNeill, Fiona. *Poor Women in Shakespeare.* Cambridge: Cambridge University Press, 2007.

Michalski, Stefan. "The Ideal Climate, Risk Management, the ASHRAE Chapter, Proofed Fluctuations, and Towards a Full Risk Analysis Model." Discussion paper for the Getty Conservation Institute: Alternative Climate Controls for Historic Buildings, Tenerife, Spain, April 2007. http://www.getty.edu/conservation/our_projects/science/climate/paper_michalski.pdf.

Middleton, Bernard C. "Gluing the Spine." In *A History of English Craft Bookbinding Technique.* London: The Holland Press, 1963.

Milgate, W., ed. John Donne, *The Epithalamions, Anniversaries, and Epicedes.* Oxford: Oxford University Press, 1978.

Mitchner, Robert W. "Wynkyn de Worde's Use of the Plimpton Manuscript of *De Proprietatibus Rerum.*" *The Library* 6, no. 1 (June 1951): 7–18.

Monro, Alexander. *The Paper Trail: An Unexpected History of a Revolutionary Invention.* New York: Knopf, 2016. Reprint, New York: Vintage, 2017.

Mosley, James. "The Technologies of Print." In *The Book: A Global History*, ed. Michael F. Suarez and H. R. Woudhuysen, 161–99. Oxford: Oxford University Press, 2013.

Muir, Bernard J. "Editing the Exeter Book: A Progress Report." In *Medieval Texts and Images: Studies of Manuscripts from the Middle Ages*, ed. Margaret M. Manion and Bernard J. Muir, 149–76. Chur: Harwood Academic, 1991.

Müller, Lothar. *White Magic: The Age of Paper.* Cambridge: Polity, 2015.

Munsch, Guy. "Condition Report for the Folger Shakespeare Library." July 2002.

Munsell, Joel. *Chronology of the Origin and Progress of Paper and Paper-Making.* 5th ed., with additions. Albany: J. Munsell, 1876.

Munsell, Joel. *A Chronology of Papermaking.* Albany: J. Munsell, 1856.

Nickell, Joe. *Pen, Ink, & Evidence: A Study of Writing and Writing Materials for the Penman, Collector, and Document Detective.* New Castle, DE: Oak Knoll Press, 2000.

Nicosia, Marissa. "To Make Lemmon Cakes." Cooking in the Archives. March 26, 2015. https://rarecooking.com/2015/03/26/to-make-lemmon-cakes.

Nixon, Rob. *Slow Violence and the Environmentalism of the Poor*. Cambridge, MA: Harvard University Press, 2011

Norton, Charles Eliot, ed. *The Love Poems of John Donne*. Boston: Houghton, Mifflin, 1905.

Orme, Nicholas. "Horman, William (1457–1535)." *Oxford Dictionary of National Biography*. Accessed June 26, 2017. http://www.oxforddnb.com/view/article/13779.

Peltonen, Markku. "Bacon, Francis, Viscount St Alban (1561–1626), lord chancellor, politician, and philosopher." *Oxford Dictionary of National Biography*. Accessed August 26, 2016. http://www.oxforddnb.com/view/article/990.

Pettegree, Andrew. *The Book in the Renaissance*. New Haven: Yale University Press, 2011.

Pettegree, Andrew. "The Legion of the Lost: Recovering the Lost Books of Early Modern Europe." In *Lost Books: Reconstructing the Print World of Pre-Industrial Europe*, ed. Andrew Pettegree and Flavia Bruni, 1–27. Leiden: Brill, 2016.

Pettet, E. C. *Of Paradise and Light: A Study of Vaughan's "Silex Scintillans."* Cambridge: Cambridge University Press, 1960.

Pickwoad, Nicholas. "Onward and Downward: How Binders Coped with the Printing Press Before 1800." In *A Millennium of the Book: Production, Design, and Illustration in Manuscript and Print, 900–1900*, ed. Robin Myers and Michael Harris, 61–106. New Castle, DE: Oak Knoll Press, 1994.

Pinzari, Flavia, Giovanna Pasquariello, and Antonella De Mico. "Biodeterioration of Paper: A SEM Study of Fungal Spoilage Reproduced Under Controlled Conditions." *Macromolecules in Cultural Heritage* 238, no. 1 (April 2006): 57–66.

Potter, Lois. "The Second Tetralogy: Performance as Interpretation." In *A Companion to Shakespeare's Works*, ed. Richard Dutton and Jean Howard, 2:287–307. Malden, MA: Blackwell, 2003.

Price, Leah. *How to Do Things with Books in Victorian Britain*. Princeton: Princeton University Press, 2012.

Prince, Tracy J. *Culture Wars in British Literature: Multiculturalism and National Identity*. Jefferson, NC: McFarland, 2012.

Prineas, Matthew. "The Dream of the Book and the Poetry of Failure in Henry Vaughan's *Silex Scintillans*." *English Literary Renaissance* 26, no. 2 (1996): 333–55.

Proot, Goran, and Leo Egghe. "Estimating Editions on the Basis of Survivals: Printed Programmes of Jesuit Plays in *Provincia Flandro-Belgica* before 1773, with a Note on the 'Book Historical Law.'" *PBSA* 102, no. 2 (June 2008): 149–74.

Pugliatti, Paola. "'More than history can pattern': The Jack Cade Rebellion in Shakespeare's *Henry VI*." *Journal of Medieval and Renaissance Studies* 22, no. 3 (Fall 1992): 451–78.

Reynolds, Anna. "'Such dispersive scattredness': Early Modern Encounters with Binding Waste." *Journal of the Northern Renaissance* 8 (Spring 2017): par. 1–43.

Richards, Jennifer, and Fred Schurink. "Introduction: The Textuality and Materiality of Reading in Early Modern England." *Huntington Library Quarterly* 73, no. 3 (September 2010): 345–61.

Robinson, Pamela. "The Format of Books: Books, Booklets, and Rolls." In *The Cambridge History of the Book in Britain*, vol. 2, *1100–1400*, ed. Nigel J. Morgan and Rodney M. Thomson, 39–54. Cambridge: Cambridge University Press, 2008.

Rose, Deborah Bird, Thom van Dooren, Matthew Chrulew, Stuart Cooke, Matthew Kearnes, and Emily O'Gorman. "Thinking Through the Environment, Unsettling the Humanities." *Environmental Humanities* 1, no. 1 (2012): 1–5.

Rudrum, Alan, ed. *Henry Vaughan: The Complete Poems*. 1976. Reprint, New Haven: Yale University Press, 1981.

Rudrum, A[lan] W. "Henry Vaughan's 'The Book'; A Hermetic Poem." *Journal of the Australasian Universities Language and Literature Association* 16 (1961): 161–66.

Rudrum, Alan. "Review: [untitled]." *Review of English Studies* 43, no. 171 (1992): 425–27.

Sansom, Ian. *Paper: An Elegy*. New York: William Morrow, 2013. Reprint, New York: William Morrow, 2015.

"Scrutinizing Surfaces." Special Issue. *Journal of the Northern Renaissance* 8 (Spring 2017). https://www.northernrenaissance.org/issues/issue-8–2017.

Senchyne, Jonathan. "Paper Nationalism: Material Textuality and Communal Affiliation in Early America." *Book History* 19 (2016): 66–85.

Senchyne, Jonathan. "Rags Make Paper, Paper Makes Money: Material Texts and Metaphors of Capital." *Technology and Culture* 58, no. 2 (April 2017): 545–55.

Serres, Michael. *The Natural Contract*. Trans. Elizabeth MacArthur and William Paulson. Ann Arbor: University of Michigan Press, 1995.

"Seventeenth-Century Letters Discovered at Knole." National Trust. Accessed July 7, 2017. https://www.nationaltrust.org.uk/knole/features/seventeenth-century-letters-discovered-at-knole.

Shammas, Carole. "How Self-Sufficient Was Early America?" *Journal of Interdisciplinary History* 13, no. 2 (1982): 247–72.

Shawcross, John T., ed. *The Complete Poetry of John Donne*. New York: New York University Press, 1968.

Sherman, William H. *Used Books: Marking Readers in Renaissance England*. Philadelphia: University of Pennsylvania Press, 2008.

Shesgreen, Sean. "Canonizing the Canonizer: A Short History of *The Norton Anthology of English Literature*." *Critical Inquiry* 35, no. 2 (2009): 293–318.

Shrivastava, Shobha. "Biodeterioration of Art Objects on Paper and Their Conservation." *Research Journal of Recent Sciences* 4 (2015): 42–46.

Simmonds, James D. *Masques of God: Form and Theme in the Poetry of Henry Vaughan*. Pittsburgh: University of Pittsburgh Press, 1972.

Simmonds, James D. "Vaughan's 'The Book': Hermetic or Meditative?" *Neophilologus* 47 (1963): 320–28.

Smith, Helen. " 'A unique instance of art': The Proliferating Surfaces of Early Modern Paper." *Journal of the Northern Renaissance* 8 (2017): par. 13–25. http://www.northernrenaissance.org/?p=4847.

Smyth, Adam. " 'Art Reflexive': The Poetry, Sermons, and Drama of William Strode (1601?–1645)." *Studies in Philology* 103, no. 4 (Fall 2006): 436–64.

Smyth, Adam. "Burning to Read: Ben Jonson's Library Fire of 1623." In *Book Destruction from the Medieval to the Contemporary*, ed. Adam Smyth and Gill Partington, 34–54. New York: Palgrave Macmillan, 2014.

Spedding, James, ed. *The Works of Francis Bacon*. Vol. 14. London: Longmans, Green, Reader, and Dyer, 1874.

Spufford, Margaret. *Small Books and Pleasant Histories: Popular Fiction and Its Readership in Seventeenth-Century England*. Cambridge: Cambridge University Press, 1981.

Stallybrass, Peter. "Books and Scrolls: Navigating the Bible." In *Books and Readers in Early Modern England*, ed. Jennifer Andersen, Elizabeth Sauer, and Stephen Orgel, 42–79. Philadelphia: University of Pennsylvania Press, 2001.

Stallybrass, Peter. "Introduction to 360 Analysis." Presented at Book 360 Symposium, Weston Library, Bodleian, Oxford, May 11–12, 2017. See https://sites.google.com/palaeome.org/lvcas-360/home.

Stallybrass, Peter, Roger Chartier, John Franklin Mowery, and Heather Wolfe. "Hamlet's Tables and the Technologies of Writing in Renaissance England." *Shakespeare Quarterly* 55, no. 4 (Winter 2004): 379–419.

Staniforth, Sarah. "Slow Conservation." *Studies in Conservation* 55, no. 2 (2010): 74–80.

Stephens, Brent. "What Have We Learned About the Microbiomes of Indoor Environments?" *mSystems* 1, no. 4 (July–August 2016): 1–9.

Stevenson, Allan. *The Problem of the Missale Speciale*. London: The Bibliographical Society, 1967.

Stewart, Alan. *Shakespeare's Letters*. Oxford: Oxford University Press, 2008.

Stoicheff, Peter. "Materials and Meanings." In *The Cambridge Companion to the History of the Book*, ed. Leslie Howsam, 73–89. Cambridge: Cambridge University Press, 2014.

Storey, H. Wayne. "Doubting Petrarca's Last Words: Erasure in MS Vaticano Latino 3195." In *Petrarch and the Textual Origins of Interpretation*, ed. Teodolinda Barolini and H. Wayne Storey, 67–91. Boston: Brill, 2007.

Strasser, Susan. *Waste and Want: A Social History of Trash*. New York: Metropolitan Books, 2000.

Stringer, Gary A., ed. *The Variorum Edition of the Poetry of John Donne*. Vols. 2 and 8. Bloomington: Indiana University Press, 1995, 2000.

Strong, James. *The New Strong's Exhaustive Concordance of the Bible*. Cambridge, ON: Thomas Nelson, 1984.

Stubbs, John. *John Donne: The Reformed Soul; A Biography*. New York: W. W. Norton, 2008.

Suarez, Michael F., and H. R. Woodhuysen, eds. *The Book: A Global History*. Oxford: Oxford University Press, 2013.

Summit, Jennifer. *Memory's Library: Medieval Books in Early Modern England*. Chicago: University of Chicago Press, 2008.

Szirmai, J. A. *The Archaeology of Medieval Bookbinding*. Brookfield, VT: Ashgate, 1999.

Tanselle, G. Thomas. *Bibliographical Analysis: A Historical Introduction*. Cambridge: Cambridge University Press, 2009.

Thomas, J. H. "The Company of White Paper Makers in Hampshire: An Inventory of Plant." *Post-Medieval Archaeology* 11, no. 1 (January 1977): 22–35.

Thomas, Marcel. "Manuscripts." In *The Coming of the Book: The Impact of Printing, 1450–1800*. Ed. Geoffrey Nowell-Smith and David Wooton. Trans. David Gerard. London: Verso, 1997.

Twyman, Michael. *The British Library Guide to Printing*. Toronto: University of Toronto Press, 1999.

Vareschi, Mark. *Everywhere and Nowhere: Anonymity and Mediation in Eighteenth-Century Britain*. Minneapolis: University of Minnesota Press, 2018.

Walker, Jonathan. "Reading Materiality: The Literary Critical Treatment of Physical Texts." *Renaissance Drama* 41, nos. 1–2 (Fall 2013): 199–232.

Wall, Wendy. "Reading for the Blot: Textual Desire in Early Modern England." In *Reading and Writing in Shakespeare*, ed. David M. Bergeron, 131–59. Newark: University of Delaware Press, 1996.

Wall-Randell, Sarah. "Marlowe's Lucan: Winding Sheets and Scattered Leaves." In *Christopher Marlowe, Theatrical Commerce, and the Book* Trade, ed. Kirk Melnikoff and Roslyn L. Knutson, 11–25. Cambridge: Cambridge University Press, 2018.

Wall-Randell, Sarah. "Marlowe, Shakespeare, and the Dead in Sheets." Paper presented at "Posthumous Shakespeare" panel, Shakespeare Association of America Conference, New Orleans, March 2016.

Waller, Gary. *English Poetry of the Sixteenth Century*. 2nd ed. New York: Longman, 1993.

Walsby, Malcolm. *The Printed Book in Brittany, 1484–1600*. Leiden: Brill, 2011.

Watson, Robert N. *Back to Nature: The Green and the Real in the Late Renaissance*. Philadelphia: University of Pennsylvania Press, 2005.

Weaver, Martin E., with F. G. Matero. *Conserving Buildings: Guide to Techniques and Materials*. New York: Wiley, 1993.

Weber, Therese. *The Language of Paper: A History of 2000 Years*. Bangkok, Thailand: Orchid Press, 2007.

Weeks, Lyman Horace. *A History of Paper-Manufacturing in the United States, 1690–1916*. Long Island: Lockwood Trade Journal, 1916.

Weis, René. "The Texts of 2 *Henry IV*." In *The Oxford Shakespeare: Henry IV Part Two*, by William Shakespeare, ed. René Weis, 78–100. Oxford: Oxford University Press, 1997.

Werner, Sarah. *Studying Early Printed Books, 1450–1800: A Practical Guide*. Hoboken, NJ: Wiley-Blackwell, 2019.

"What Is the Office of the President, and of All Those Who Sit There Officially: As to the Scribe of the Treasurer." In *The Dialogue Concerning the Exchequer, circa 1180*. From *The Avalon Project: Documents in Law, History, and Diplomacy*, Yale Law School Lillian Goldman Law Library. Accessed December 6, 2018. http://avalon.law.yale.edu/medieval/excheq.asp.

Whitver, H. Austin. "Materiality of Memory in Shakespeare's Second Tetralogy." *Studies in English Literature* 56, no. 2 (Spring 2016): 285–306.

Williamson, George. *Six Metaphysical Poets: A Reader's Guide*. Syracuse: Syracuse University Press, 1967.

Wolfe, Heather. "How Expensive Was Writing Paper in Early Modern England?" Paper presented at Renaissance Society of America Conference, Washington, DC, March 22–24, 2012.

Wolfe, Heather. "Was Early Modern Writing Paper Expensive?" *The Collation*. February 13, 2018. https://collation.folger.edu/2018/02/writing-paper-expensive.

Yates, Julian. *Error, Misuse, Failure: Object Lessons from the English Renaissance*. Minneapolis: University of Minnesota Press, 2003.

CATALOGUE ENTRIES

Catalogue entry for Folger call number S2915 Fo.4 no.18. *Hamnet: Folger Library Catalogue*. Accessed December 10, 2018. http://hamnet.folger.edu/cgi-bin/Pwebrecon.cgi?BBID=108071. Shakespeare et al., *Mr. William Shakespear's Comedies, Histories, and Tragedies: Published According to the True*

Original Copies: Unto Which Is Added, Seven Plays, Never before Printed in Folio, 4th ed., Wing S2915. London: Printed for H. Herringman, 1685.

Catalogue entry for Folger shelf mark STC 22274 Fo.2 no.07. *Hamnet: Folger Library Catalogue*. Accessed December 6, 2018. hamnet.folger.edu.

Catalogue entry for Folger shelf mark STC 22288. *Hamnet: Folger Library Catalogue*, Accessed December 6, 2018. hamnet.folger.edu. William Shakespeare et al., *The Second Part of Henrie the Fourth*. STC 22288. London: Printed by V[alentine]. S[immes] for Andrew Wise, and William Aspley, 1600.

Catalogue entry for Huntington call number 497071. *Huntington Library Catalogue*. Accessed December 10, 2018. https://catalog.huntington.org/record=b1066975. Edward Coke et al., *Le tierce part des Reportes del Edward Coke l'attorney generall le Roigne* [. . .], STC 5499.2. [London: Printed by Adam Islip] in aedibus Thomae Wight, 1602.

Catalogue entry for University of Texas at Austin Library call number PFORZ Add. 22. *University of Texas Libraries, Library Catalog*. http://catalog.lib.utexas.edu/record=b3444254~S18. John Donne, John Marriot, and Richard Marriot, *Poems*, Wing D1869. London: Printed for John Marriot [. . .] and are to be sold by Richard Marriot, 1650.

INDEX

Note: Page numbers in italic type indicate illustrations.

ACKNOWLEDGMENTS

You know who you are. Mentors, teachers, colleagues, inquisitive bibliophiles, and generous skeptics. Librarians, conservators, artisans, registrars, teaching assistants, research assistants, students, hiking companions, Adirondackers, Delawareans, Sconnies, musicians, farmers, land surveyors, dogs and one cat, neighbors, friends. Daughter, sister, brother, mother, father. Looking closely at the heterogenous flecks and fibers embedded in handmade paper, we begin to see the influence of other story lines. Looking back at this work on the nature of early handmade paper, I find myself similarly drawn to notice and appreciate the many intersecting lives and story lines that have given energy, detail, complexity, and texture to the pages of this book.

This work was sustained by the optimism of various institutions that supported and funded my research. I am particularly grateful for a short-term fellowship at the Folger Shakespeare Library, summer funding from the Graduate School at the University of Wisconsin–Madison, a semester-long fellowship at the Institute for Research in the Humanities at the University of Wisconsin–Madison, and a semester-long fellowship at McGill University's Institute for the Public Life of Arts and Ideas (where Paul Yachnin, Fiona Ritchie, Katie Zien, and Sheetal Lodhia generously engaged with and improved the work). Chapter 3 was spurred forward at a critical juncture by an invitation to collaborate with Robyn Adams and Matthew Symonds in a workshop at the Centre for Editing Lives and Letters funded by the University College London (thanks also to Alexandra Ault, Sjoerd Levelt, Lucy Munro, and Tiffany Stern for generative questions). Travel funding to participate in the Beasts to Craft workshop at Oxford University further improved this book, especially Chapters 4 and 5, and I am grateful to Matthew Collins for facilitating such a promising cross-disciplinary conversation about the future of archival research. At various stages in the research and writing process, I received valuable feedback from audiences at the University of Delaware, the University of Georgia, Lawrence University, McGill University, the University of North Carolina at Charlotte, and the Workshop in the History of Material Texts at the University of Pennsylvania.

My colleagues in the Department of English at the University of Wisconsin–Madison have improved this book and enlivened the process of writing it. Mark Vareschi has been a dear friend and an invaluable source of insight on topics ranging from media history and archival research to bike maintenance and basketball. Karen Britland has been overgenerous at every turn; she read every word of a much less polished version of this book and helped me find my way out of a dozen dead ends. Christa Olson has been a candid and kind friend and colleague. At timely moments throughout the process of finishing this first book, Elizabeth Bearden, Leslie Bow, Lisa Cooper, Martin Foys, Susan Friedman, Theresa Kelley, Jesse Lee Kercheval, Mario Ortiz-Robles, and Russ Castronovo offered perspective, key insights, and sustaining encouragement. I am grateful for research assistance from Leah Misemer, Ilango Villoth, and Emily Loney. For insightful questions and new perspectives, I'm thankful to students in my spring 2017 graduate class, "Book Ecology: Then to Now, There to Here."

The Nelson Institute for Environmental Studies and its affiliated faculty have provided another intellectual home outside my home department. In fall 2013, early in the writing process, I participated in the Faculty Development Seminar "Environmentalism in the Age of the Anthropocene"; the course helped me to better understand how my book history and environmental humanities interests aligned, and I'm so grateful to Rob Nixon, who led the course, and to each of my classmates: Samer Alatout, Anna Andrzejewski, Katarzyna Beilin, Will Brockliss, Lynn Keller, Richard Keller, Gregg Mitman, Larry Nesper, Sainath Suryanarayanan, Alberto Vargas, and Lydia Zepeda. Involvement in biweekly colloquia and annual outings with the Center for Culture, History, and Environment has similarly prodded me to see productive disciplinary overlaps (and underlaps) I had not noticed before; in addition to those already mentioned above, I am grateful to Elizabeth Hennessy, Curt Meine, Heather Swan, Shari Wilcox, Cathy Middlecamp, and William Cronon. More recently, CHE's collaboration with the Rachel Carson Center in Munich, Germany, has led to meaningful, sustained exchanges with international colleagues during trips along the Mississippi and Danube rivers. I am particularly grateful to Annka Liepold, Christof Mauch, Felix Mauch, Wolfram Mauser, and Eleonora Rohland for sharing ideas and insights. This book owes much to ongoing opportunities to explore the complexities of key environmental topics while traveling, swimming, hiking, singing karaoke, sharing meals, and riding bikes with brilliant graduate students and colleagues.

At the University of Wisconsin–Madison, I have found another home in the library. In Special Collections, Susan Barribeau, Kaydian Campbell, Cynthia Lundey, David Pavelich, Robin Rider, Laura Schmidt, Carly Sentieri, Susan Stravinski, and Lisa Wettleson have made my research and teaching better and more exploratory. Tracy Honn and Jonathan Senchyne have geeked out with me and my students about the histories of papermaking and printing. In Special Collections and extending out to various sites around campus, the Holding History program has kept my archival interactions energetic and wonderfully inquisitive. I'm grateful to Kaydian Campbell, Ryan Hussey, Sam Landowski, Alex Smith, John Spoerl, and Ilango Villoth and to the other student curators who have been involved in that program, as well as to codirector Sarah Marty.

I am grateful, too, for opportunities to make and talk about paper with expert artisans: Timothy Barrett at the University of Iowa Center for the Book's Research and Production Paper Facility, John Krill at the conservation lab at Winterthur Museum, Tom Leech at Palace Press in New Mexico, Drew Matott and Margaret Sheppard in varied locales, and Carole Vossler at BluSeed Studios in the Adirondacks. Robert Possehl's papermaking expertise and his keen attention to plants and landscapes are inextricably embedded in *The Nature of the Page*, and I am grateful for his insight, his willingness to collaborate and teach together, and his friendship.

Other communities sustained me in meaningful ways that made the work possible. In the Adirondacks, I am grateful to William and Kirstin Edelglass, Erik and Dawn Klotzko, Nan and Steve Amstutz, Ruth Kuhfahl, Bill Bowman, Steve and Dee Bowen, Andrew Gushee, Tyler and Shannon Eaton, Linda and Joe Bogardus, the two librarian Karens (Glass and Rappaport), who kept me in books when I was in the woods, and to two classes of potlucking readers in my Shakespeare Out Loud community courses.

Classmates and mentors from the University of Delaware continue to inspire and challenge; thank you to Kristen Poole, Jim Beaver, Martin Brückner, Jim Dean, Hannah Eagleson, Darlene Farabee, Virginia Garnett, McKay Jenkins, Matt Kinservik, Lois Potter, Charles E. Robinson, Jeanne Murray Walker, Miranda Wilson, and Julian Yates. From the time I showed up for a masters-level seminar to now, librarians and staff at the Folger Shakespeare Library have been generous and encouraging; thanks are especially due to Erin Blake, Erik Castillo, Rachel Dankert, LuEllen DeHaven, Marta Giron, Rosalind Larry, Camille Seerattan, and Betsy Walsh (who is much missed).

Midway through my graduate career, I began taking a book history course at the University of Pennsylvania with Peter Stallybrass. In addition to all I learned and continue to learn from Peter, the course provided me with another community of inspiring friends and colleagues. I'm grateful to Margreta de Grazia, Stephanie Elsky, Lynne Farrington, Brooke Palmieri, John Pollack, and Thomas Ward for their interest in creative input at a critical stage in the development of a slowly germinating version of this book. Zachary Lesser has encouraged this work at various stages, giving me the confidence I needed to share it more broadly. Jerry Singerman guided me through the editorial process, and two anonymous readers helped me to better understand, clarify, and communicate my vision for this book.

I could never name each generous person who engaged with some form of these ideas and offered generative feedback. By asking questions and especially by asking me to explain fuzzy ideas, scholars at conferences and students in classes and friends and family in various places have all improved the quality of this work. On many occasions, colleagues not only engaged with my work, but also shared their own in advance of publication; for such generous openness, I am especially grateful to Jeff Knight, Helen Smith, Peter Stallybrass, Alan Stewart, Sarah Werner, Bill Sherman, Randy McLeod, Vin Nardizzi, and Thomas Ward.

By sharing work in progress and by offering generous, timely feedback on my developing manuscript, my writing group improved the quality of work and life during some of the most stressful seasons of this project. For that, I'm deeply grateful to Claire Bourne, Megan Heffernan, Adam Hooks, Tara Lyons, and Whitney Trettien.

In addition to those named above, numerous colleagues have generously read and offered insightful feedback on drafts of this book. For such generosity and insight, I am grateful to Katherine Acheson, Kristen Berg, Martin Foys, Alan Galey, Miriam Jacobson, Sujata Iyengar, Rebecca LaRoche, Sarah Marty, Steve Mentz, Jen Munroe, Philip Palmer, Doug Pfeiffer, Ben Robinson, Dan Shores, and a host of co-participants in a half dozen seminars at various Shakespeare Association of America meetings.

Heidi Brayman, Karen Britland, Jonathan Walker, Heather Wolfe, Virginia Garnett, Mark Vareschi, and Robert Possehl read more than a couple drafts over the years. These friends have improved the quality of my work and, in the meantime, have made the process of researching and writing more intelligent and companionate. Others inspired me most by helping me to take breaks from the work, by encouraging me to hike, dance, sing, travel, sleep, eat, garden, sled, swim, camp out, visit museums, go to concerts, binge-watch shows, cook food, and so on. Thanks to Tom Atema, Saby Cordoba, Bryan and Leah Corley, Sharon Eliashar, and Rob and Meg; to Heather, Qindi, Sylvana, and Thea; to Cafe Carpe and The Recovery Lounge; to my brother, Caleb Calhoun, and my mother, Betsi Calhoun. I owe a special debt of gratitude to my father and dear friend, Mike Calhoun whose work ethic, unconditional support, and resiliency are sustaining and inspiring. Finally, thanks and admiration to Misty Arden Calhoun, namesake of my sister. Chickadee's curiosity, thoughtfulness, capability, ardor, and joy amaze me, and I dedicate *The Nature of the Page* to her.